EARLY SCIENTIFIC COMPUTING IN BRITAIN

Early Scientific Computing in Britain

MARY CROARKEN

CLARENDON PRESS · OXFORD

This book has been printed digitally and produced in a standard specification in order to ensure its continuing availability

OXFORD
UNIVERSITY PRESS

Great Clarendon Street, Oxford OX2 6DP

Oxford University Press is a department of the University of Oxford.
It furthers the University's objective of excellence in research, scholarship,
and education by publishing worldwide in

Oxford New York

Auckland Cape Town Dar es Salaam Hong Kong Karachi
Kuala Lumpur Madrid Melbourne Mexico City Nairobi
New Delhi Shanghai Taipei Toronto
With offices in
Argentina Austria Brazil Chile Czech Republic France Greece
Guatemala Hungary Italy Japan South Korea Poland Portugal
Singapore Switzerland Thailand Turkey Ukraine Vietnam

Oxford is a registered trade mark of Oxford University Press
in the UK and in certain other countries

Published in the United States
by Oxford University Press Inc., New York

Reprinted 2011

ISBN 978-0-19-853748-9

To Julian

ACKNOWLEDGEMENTS

This work has grown out of my 1986 Ph.D. thesis 'The centralization of scientific computation in Britain 1925–1955'. The research was funded by the Science and Engineering Research Council and carried out at the Computer Science Department, University of Warwick. My thanks go to the staff at Warwick, particularly Martin Campbell-Kelly.

I wish to acknowledge the assistance of the staff at various libraries, archives, and companies which have given me access to material. These include Warwick University Library, Birmingham University Library, Birmingham Public Library, the Science Museum Library, the British Library, Computer Science Library of Glasgow University, Cambridge University Library, Churchill College Archives, Cambridge, ICL Ltd., the Royal Greenwich Observatory, the Public Records Office, the National Physical Laboratory, Department of Statistics and Computational Mathematics, Liverpool University, and the Scientific Computing Service Ltd. Also the many people who granted me personal interviews or corresponded with me. Their contributions are listed in the bibliography.

In addition I wish to acknowledge the assistance given to me by the Computer Department of Anglian Windows Ltd. for their technical help on my move from Warwick University to Norwich.

My thanks also to Oxford University Press, L. Fox and J. Howlett who have offered much appreciated constructive criticism and to Julian Cubitt for his patience.

CONTENTS

LIST OF PLATES

1

Introduction

In considering the centralization of scientific computation this book attempts to link together two areas of study: the prehistory of computers and the attitudes of scientists towards using machine methods of computation. More specifically, it examines how scientific computation was organized in Britain from the early 1900s to the mid-fifties. That is, from the time desk calculators began to become both readily available and reliable, to the invention and construction of the stored program computer. During this 50-year period the way in which computation was carried out changed significantly not only in terms of the technology used but also in the way in which computation was organized and financed. Advances in applied science meant that scientists and administrators had to find new ways of coping with an increasing demand for computation. The idea of centralizing computer power and providing user support developed slowly during the period. The concept of centralized compution culminated in 1945 with the creation of the National Physical Laboratory (NPL) Mathematics Division.

The Mathematics Division of the NPL was set up to act as a national computing resource in response to an increasing demand for fast scientific computation. It theoretically represented centralized computing in the extreme. Most of the computing technology initially installed at the NPL had been available since the mid-thirties or earlier but this was the first time such a wide selection of computing equipment had been made available to a large number of users. The creation of the NPL Mathematics Division also illustrated a recognition by scientists and civil servants that the scientific community as a whole would benefit from the expertise computing specialists could provide.

When the concept of the stored program computer began to emerge after the war, the NPL Mathematics Division was in an ideal position, in principle, to exploit these new ideas and build a machine to serve the scientific community in Britain. The NPL was not, however, alone in developing computers in Britain; computer projects were also initiated at Cambridge and Manchester Universities in the late 1940s. The existence of powerful computing resources elsewhere in Britain meant that the NPL Mathematics Division lost its status as *the* computing centre in Britain almost as soon as it had been established.

The evolution of scientific computing centres was a gradual one which stemmed from the work of scientists throughout the early twentieth century at a variety of institutions. To provide a framework in which to discuss this evolution nine characteristics which help to describe a computing centre have been identified.

1. *Users:* What group of people did the centre serve and how easy was it for them to access the centre?
2. *Personnel:* How large a staff did the centre support? Did the staff consist of both trained computing staff and experienced mathematicians?
3. *Machinery:* What computing machinery did the centre have access to?
4. *Computing activity:* What type of computations were carried out? Did they cover a wide range of computing techniques and applications?
5. *Advisory role:* Did the centre provide an advisory service to users including guidance during the early stages of an investigation as to the best way to approach a problem?
6. *Numerical research:* Did the centre undertake research into numerical techniques either independently or on behalf of its users?
7. *Computing machinery research:* Did the centre undertake research into new, or improved, types of computing machinery?
8. *Publications:* Did the centre publish the results of its work or in some way attempt to educate its users or disseminate information about the latest computing techniques?
9. *Library facilities:* Did the centre maintain a specialist library which its users could consult?

These characteristics can be identified in varying degrees in the computing centres discussed in later chapters. Very few of the centres considered could have been described as possessing all nine characteristics but the questions they pose can be used to illustrate changes in the way in which computation was carried out during the period from 1900 to 1950.

Although it is useful to use the characteristics above to help assess to what degree a specific institution represents a computing centre, they are in no way intended to be judgemental. In many cases the institution under discussion neither wanted to be nor claimed to provide a comprehensive computing service. The characteristics have been used purely to illustrate the emergence of scientific computing centres over a period of time.

An important distinction is made between scientific and commercial computation. The term scientific computation is used to describe the very varied types of numerical calculation which applied scientists, engineers, surveyors, and others carry out in the course of their work. Commercial computation includes all accounting, payroll, and stock control work carried out by businesses and others. The following chapters concentrate solely on scientific computation. A second distinction needs to be drawn in order to define the scope of this study. Computation is taken to mean numerical calculations carried out in order to achieve a solution for a practical end.

Although research into numerical methods is considered to be a very important aspect of the work of a computing centre, small groups working independently on some aspect of numerical analysis have not been considered as computing centres. Such groups often made a very great contribution to the history of numerical analysis which is too vast a subject to be included here.

Before looking at the computing centres which began to emerge in the early twentieth century, it is important to have some idea of the type of computational aids which were available at that time and to briefly consider their history.

2

Background: aids to computation

The most commonly used computational aids available in the 1900s were mathematical tables, slide rules, nomograms, mathematical instruments, and desk calculating machines. Of these only the mathematical tables and the desk calculators lent themselves to the type of large-scale repetitive calculations which could be centralized with advantage. The increasing use of desk calculating machines during the early twentieth century laid the foundation for computing centres to become a possibility but nevertheless mathematical tables, slide rules, and mathematical instruments had an important role to play.

MATHEMATICAL TABLES

With the exception of the abacus, mathematical tables are the oldest and most well-established aid to computation. Fundamentally the purpose of a mathematical table is to provide precomputed values which can be used as part of a large and/or complex calculation. By eliminating some part of a calculation mathematical tables save time and labour in computational work. Glashier (1911*b*) stated that

> The intrinsic value of a table may be estimated by the actual amount of time saved by consulting it: for examples, a table of square roots to ten decimals is more valuable than a table of squares, as the extraction of the root would occupy more time than than the multiplication of the number by itself. The value of a table does not depend upon the difficulty of calculating it; for, once made, it is made for ever, and as far as the user is concerned the amount of labour devoted to its original construction is immaterial. (p.325)

The main classes of mathematical tables are arithmetical, natural trigonometrical functions, logarithms to base 10 (to include logarithms of numbers and logarithms of the trigonometrical functions), logarithms to base e, algebraic constants (for example, Binomial coefficients, Bernoulli numbers, functions of π, e, etc.), and transcendental functions.

Published arithmetical tables have been available since the early seventeenth century. In 1610, Herwart ab Hobenburg produced a multiplication table giving products from 2×1 to 1000×1000. Multiplication tables were also published throughout the eighteenth and nineteenth centuries. The most well-known was Crelle's *Rechentaflen* first published in 1820, which gave products up to 1000 × 1000. Crelle's *Rechentaflen* was considered to be one of the most convenient set of multiplication tables of its era and was published in a number of subsequent editions in many languages. Crelle's popularity continued well into the twentieth century to such an extent that in 1924 E. T. Whittaker, in his preface to *The calculus of observations*,[1] listed Crelle's *Calculating Tables* as an essential tool for anyone undertaking numerical computation.

Tables of quarter squares were also used to facilitiate multiplication. By using the relation $ab = 1/4(a+b)^2 - 1/4(a-b)^2$ with a table of quarter squares the process of multiplication could be reduced to simple subtraction. According to Glashier (1911*b*) the earliest known table of quarter squares was published by A. Voisin in 1817. However, the largest such table, and most frequently used in the late nineteenth and early twentieth centuries, was J. Blater's *Table of quarter-squares of all whole numbers from 1 to 200 000* (1888).

Other common types of arithmetical table were tables of squares, cubes, square roots, cube roots, and reciprocals. Glashier (1873) states that the origin of these tables dates back to the fourteenth century. By far the most widely used in the late eighteenth and early nineteenth centuries was Barlow's *New mathematical tables* first published in 1814. A second edition was published in 1840 by De Morgan which superseded the original. In 1930 L. J. Comrie produced a third edition; a fourth edition appeared in 1941 also edited by Comrie.[2]

Tables giving the natural values of the trigonometrical functions were also published from an early date. Glashier (1873) credits P. Apian with the first published table of natural values of the trigonometrical functions in 1533 but does not supply any bibliographic details to support this claim. The most important of the early tables of trigonometrical functions was Rheticus's *Opus Palatinum* published in 1596 which gave values of all six trigonometrical functions to 10 figures for every ten seconds of arc. After Rheticus's death Pitiscus published, in 1613, his *Thesaurus* as a new edition of Rheticus's work which took the table to 15 figures and corrected some of the errors found in the original table.

The publication of John Napier's *Descriptio* in 1614, only a year after Pitiscus's *Thesaurus* appeared, effectively put an end to the use of tables of natural values of trigonometrical functions. The *Descriptio* announced Napier's invention of logarithms and gave, as an example, a table of logarithms of sines. Napier's original logarithms were to base e^{-1} and had been devised as a means of reducing the process of multiplication and division to simple addition and subtraction respectively. The *Descriptio* had an almost immediate effect on the natural philosophers and mathematicians of the seventeenth century who quickly adopted logarithms as the standard means of computation.

Soon after Napier's *Descriptio* had been published Henry Briggs, Professor of Geometry at Gresham College, London, approached Napier with the suggestion that logarithms to base 10 would be more generally useful.[3] With Napier's approval Briggs published logarithms to base 10 of numbers from 1 to 1000 to 14 figures in 1617. He continued to work on calculating logarithms to base 10 and in 1624 published his *Arithmetica logarithmica* giving tables of logarithms of numbers from 1 to 20 000 and from 90 000 to 100 000 to 14 figures. Briggs's work was followed in 1628 by Adrian Vlacq's *Arithmetica logarithmica, sive logarithmorum chiliades* . . . which gave logarithms to base 10 of numbers from 1 to 100 000. Vlacq also included tables which gave logarithmic values of sines, tangents, and secants. Vlacq called his work a second edition of Briggs's *Arithmetica logarithmica* but had himself calculated the logarithms of numbers from 20 000 to 90 000. It is from Briggs's and Vlacq's tables (with corrected errata) that many subsequent tables of logarithms have been derived.

In the early twentieth century there were many tables of logarithms available. In the range of 7 figure tables Babbage's *Tables of logarithms of the natural numbers from 1 to 108,000* was considered by Bell and Milne (1914) to be the most accurate. But tables giving 7-figure logarithms up to 108 000 were also available in later editions of Callet's *Tables portavives* . . . and Chambers' *Mathematical tables* which were both much used collections. Logarithms to 8 figures were given by Bausinger and Peters (1910) and to 10 figures by Duffield (1895–6). Tables giving logarithms up to 27 figures had also been published, for example by Thoman (1867), but the need for anyone to work to that degree of accuracy was rare.

As with logarithms of numbers, many tables of logarithms of the trigonometrical functions had been published by the early twentieth century. The most celebrated in terms of accuracy was Andoyer's *Nouvelle tables trigonometriques fondamentales (logarithms)* published in 1911. These gave logarithmic values of the trigonometrical functions to 14 figures for every tenth second. Tables giving logarithms of trigonometrical functions to 7 places were easily accessible in collections of mathematical tables such as Chambers.

The most extensive tables of logarithms to base *e* available in the early twentieth century were those calculated by J. Wolfram and published in Vega's *Thesaurus logarithmorum completus* . . . (1794)[4] which gave logarithms for numbers from 1 to 22 000 and logarithms of primes up to 10 009 both to 48 figures. Other tables giving logarithms to base *e* were Barlow's *Tables* giving tables to 7 figures, Hantschl's *Logarithmisch-trigonometrisches handbuch* (1827) giving tables to 8 figures and J. Saloman (1827) giving tables to 10 places.

Tables giving Binomial coefficients, Bernoulli numbers, functions of e, π, etc., Gamma, Legendre, and Bessel functions were also available and are listed in order of usefulness in Bell and Milne's '*A working list of mathematical tables*' (1914). In addition specialist tables to assist seamen, surveyors, astronomers, engineers, and others were published either individually, in collections of tables (for example, in the Chambers 1878 edition), or in specialist journals.

Mathematical tables were, therefore, a very well-established and common aid to computation easily available through book sellers. Tables giving values to many figures were available for those who needed them but tables to 7 figures or less were published for every day use.

SLIDE RULES

The *Encylopaedia Britannica* gives the definition of a slide rule as a device

> consisting of graduated scales capable of relative movement, by means of which simple calculations may be carried out mechanically. In ordinary slide rules these operations include multiplication, division and extraction of square roots, as well as, in some cases, calculation of trigonometrical functions and logarithms. (Keuffel 1971)

As such the slide rule is an aid to computation.

Slide rules were invented in the early to mid-seventeenth century as a direct result of Napier's work on logarithms.[5] In 1620 Edmund Gunter constructed a straight rule bearing a logarithmic scale. He used a pair of dividers to carry out multiplication and division on the scale. The invention, around 1630, of the slide rule as we understand it today, where the dividers are replaced by sliding rules, has often been credited to William Oughtred.[6] However, Richard Delamain published an independent description of a circular slide rule in 1630 which puts Oughtred's claim in some dispute (Williams, M. R. 1982). What is clear, however, is that slide rules were being continually developed and improved throughout the seventeenth century by Oughtred, Delamain, Thomas Brown, Brissaker, John Brown, Coggeshall, Everard, and others. Slide rules were relatively simple and cheap to produce, and were commercially manufactured by English instrument makers throughout the seventeenth and eighteenth centuries.

Slide rules continued to improve during the nineteenth century by which time many different types were available. The standard slide rule of the late nineteenth century was based on a 1859 design by Mannheim which incorporated a cursor to assist in the reading of the scales (Plate 1). The standard slide rule consisted of four scales, A, B, C, and D. The top two scales, A and B, were logarithmic scales used for multiplication and division. Scale C was identical to the A and B scales but was used for finding factors. The fourth scale D differed and was used to find squares and roots. In some slide rules the back of one or more of the slides was graduated in such a way as to provide a table of sines, tangents, or logarithms.

Slide rules were used by a wide range of people in the seventeenth, eighteenth and nineteenth centuries. The standard rule could be used by anyone who required a means of doing basic calculations quickly to an accuracy of approximately 2 or 3 figures where the number of repeated calculations was relatively small. For engineers, surveyors and others the slide rule was a cheap and powerful tool which had the added advantage of being easily portable. In addition many specialist rules were available. For example, there were slide rules designed specifically for use by Excise Officers, engineer's rules which carried a variety of constants, rules for calculating interest and performing related calculations, rules for calculating the weight and girth of cattle, nautical slide rules, carpenters rules, rules for paper manufacturers, and many others. In Baxendale's *Calculating machines and instruments: catalogue of the collections in the Science Museum* (1926) 103 different types of slide rule are identified.

Although the slide rule was a commercially produced and popular device, particularly in Britain, it was not used universally. In an article in 1885 in the popular journal *Nature* C. V. Boys called for more people of the 'calculating community' to use the slide rule. Boys identified two common misconceptions which prevented scientists and engineers of the late nineteenth century using the slide rule in their work. The first of these misconceptions was that because slide rules were based on logarithms their use must be complicated; the second was that slide rules gave only approximate results. Boys dealt with both of these objections and also gave examples where the use of a slide rule would be inappropriate; for example, the analysis of data presented to 4 or more significant figures (Boys 1885).

The point that not everyone was familiar with the use of the slide rule was also made by Stokes in 1914. Stokes did not assume that his reader had a prior knowledge of slide rules and went to some length to explain the principles of their operation.

Slide rules were, therefore, a readily available aid to calculation at the beginning of the twentieth century and, although not universally employed, they were fairly common.

NOMOGRAMS

Another form of calculating tool available to scientists at the beginning of the twentieth century was the nomogram. Nomograms are graphs which represent a solution for a particular class of problem. A nomogram can be difficult to construct but once it has been prepared it is relatively simple to read off the required data either as a point on the nomogram or as a line. Nomograms were made popular, particularly in France, in the late nineteenth century by d'Ocagne but some earlier examples have been found. For example, Evesham (1985) draws attention to Lalanne's 'Universal calculator' published in England in the 1840s. This nomogram could be used for multiplication, division, calculating integer powers, extracting roots, and calculating the areas of circles and the volumes of spheres. The applications of Lalanne's 'Universal calculator' were therefore very similar to those of the slide rule.

Nomograms were used principally by engineers, land surveyors, and the military. Specialist nomograms for particular applications, such as trajectory calculations, were readily available. Once the nomogram had been constructed and published, it was an easily portable, simple, and accurate computing tool. Nomograms were more popular in continental Europe than in Britain where they were overshadowed by the slide rule.

MATHEMATICAL INSTRUMENTS

During the late nineteenth century a new class of computational aid emerged; the mathematical instrument. In these instruments numbers were represented by physical quantities such as the length of a rod, the angular rotation of a wheel, the level of a liquid, or a voltage. Their accuracy was therefore limited by the precision to which measurements could be taken. A slide rule could, therefore, be classed as a mathematical instrument. Mathematical instruments were often built to perform a specific function, for example integration or harmonic analysis. Today such instruments are generally called analogue machines.[7]

The most common form of mathematical instrument available in the nineteenth century was the planimeter. Planimeters were used to measure the area of a closed curve; the curve might be obtained from a scaled drawing or a plotted function. Planimeters were used chiefly by engineers to calculate cross-sectional areas which would otherwise have to be found numerically using Simpson's rule. Although the invention of the planimeter is attributed to Hermann in 1814 (Baxendale 1929*b*, Carse and Urquhart 1914, Bromley 1984), the first planimeter to be commercially manufactured was designed by Oppikofer in 1826 and sold by Ernst of Paris. Oppikofer's planimeter used a pointer to trace the boundary of the closed curve under examination. As the pointer traced the boundary a cone rotated in response to movement of the pointer along the x-axis. Perpendicular to, and in contact with, the surface of the cone rolled a wheel. The distance of the wheel from the apex of the cone varied with the movement of the pointer along the y-axis. The rotation of the wheel, which was recorded by a counter, was proportional to the area enclosed by the curve.

Throughout the nineteenth century planimeters were manufactured and sold in increasing numbers as their accuracy improved. One important change was the replacement of the integrating cone by a circular disk by Welti in 1849 which allowed the integration of negatively valued functions. In 1856 J. Amsler designed the polar planimeter (Plate 2) 'which in consequence of its simplicity, handiness in use, and low price, soon drove all the older forms which, when well made were necessarily expensive, out of the field' (Henrici 1894, p.508). The polar planimeter used a method in which the integrating wheel used both rolling and slipping movements to record the area traced by the pointer. By 1884 over 12 000 Amsler planimeters had been sold (Baxendale 1929*b*) and the instrument continued to be widely used during the late nineteenth and early twentieth centuries.

Because of variations between the theoretical and actual construction planimeters slight errors were introduced. The error in the results given by Amsler planimeters could not be adequately determined and hence in some cases Amsler planimeters were not sufficiently accurate. In 1880 Coradi, a firm of instrument makers based in Zurich, began to produce the compensating planimeter which measured a given area from two different positions allowing the error in the result to be significantly reduced.

Planimeters were manufactured by several companies besides Amsler and Coradi. For example A. Ott of Kempton in Bavaria, Starke of Vienne, Welti of Zurich, and Richard Frères of Paris all manufactured and sold planimeters. In addition to the commercially available planimeters used by engineers, many specialized planimeters were constructed by individual scientists but were not widely used.

Another common form of analogue calculating machine used in the late 1800s and early 1900s was the integrator. Integrators, like planimeters, measured the area of a given closed curve but also gave $\int y^2 \,.\, dx$ and $\int y^3 \,.\, dx$ of the curve $y=f(x)$. They were used mainly by engineers and designers to compute centres of gravity, sheer forces, bending moments, and for stability calculations. The most widely available forms of integrator were manufactured first by Amsler and later by Coradi. The machines were sold throughout Europe.

Mathematical instruments which solved linear differential equations by giving the integral curve were called integraphs. The most well-known integraph was manufactured by Coradi and based on an 1878 design by Abdenk-Abakanowitz (Plate 3). Several other types of integraph were constructed by individual scientists but these were usually purpose-built machines and not commercially sold. Although integraphs are described in the literature alongside planimeters and integrators (d'Ocagne 1928, Baxendale 1929*b*) they were not used on nearly so large a scale. In 1914 A. M. Robb, of the Department of Naval Architecture at Glasgow University, observed that while planimeters and integraphs were in daily use in British shipyards integraphs were not. Robb thought that there were 'probably only three or four [integraphs] in this country (p.207).

Another important, although not commonly used, class of mathematical instruments were the harmonic analysers. Harmonic analysers were used to determine the Fourier coefficients of a periodic production expressed as a curve. Sir William Thompson (Lord Kelvin) designed and built the first harmonic analyser in 1876 (Plate 4).

Thomson was a member of the British Association Tides Committee which had been set up in 1867 to ensure that the necessary harmonic analyses on collected tidal data was performed in order that future tides in ports around the world could be predicted. The work had been carried out using arithmetical and graphical methods and had been funded by the British Association, the Royal Society, and the Indian government. To speed up the process Thomson used eleven interconnected integrating mechanisms, designed by his brother James (Thomson and Tait 1890), to mechanically perform the harmonic analysis of a tidal curve and hence he significantly reduced the amount of labour required for this type of work.

Unlike planimeters and integrators, harmonic analysers were large, cumbersome, and complex devices that were not easily portable. Several other harmonic analysers were built by individual scientists for their own use such as Henrici, Yule, Michealson and Stratton, and Boucherot. Henrici's machine was subsequently improved and manufactured on a limited scale by Coradi. However, the high cost of building such complex machines and the many numerical and graphical methods of harmonic analysis available during the late nineteenth century goes some way towards explaining why harmonic analysers were not more common.

A further class of mathematical instruments built during the nineteenth and early twentieth century were the algebraic equation solvers. Like harmonic analysers, equation solvers were not usually commercially manufactured and tended to be built by individuals. Reviews of early equation solvers are given by Jacob (1911) and Frame (1945); these illustrate the wide variety of different types of machine which were built. The majority of designs for machines to solve algebraic equations were based on systems of mechanical linkages but others used pulleys and liquid level balancing mechanisms. Baxendale, curator of scientific instruments at the Science Museum, South Kensington, was dismissive of equation solvers as a whole. In his review of mathematical instruments for the 1929 *Encyclopaedia Britannica* Baxendale wrote,

> Many instruments have been designed for the mechanical solution of algebraic equation. These are in general more remarkable for the ingenuity displayed in their design than in the practical value of the results obtainable. No instrument of this type has been brought into extensive use, but a few of them have found a limited application in cases where a considerable number of approximate results are required. (Baxendale 1929*b*, p.69)

However, the importance of mathematical instruments at the beginning of the twentieth century is illustrated in the *Handbook of the Napier tercentenary celebrations* which was published not only to commemorate Napier's tercentenary but was also intended to provide a review of up-to-date methods of computation (Horsburgh 1914). Almost one-third of this important book is given to describing the use and design of a wide range of such instruments. It is interesting to note that many of the exhibits discussed in the section were contributed by the Department of Natural Philosophy at Edinburgh University and not the Mathematical Laboratory which had contributed desk machines, computing forms, and mathematical models. While physical scientists were interested in devising instruments to help them with specific aspects of their work, mathematicians used more general tools such as logarithms and desk calculators.

DESK CALCULATING MACHINES

Whereas slide rules and mathematical instruments went into popular use relatively soon after their invention in the 1630s desk calculating machines did not. Calculating machines first began to appear in the mid-1600s but although many machines had been designed and built since Pascal's machine of 1642,[8] the first commercially manufactured calculating machine was designed by Thomas de Colmar in 1820 and called the 'Arithmometer'.

Thomas's Arithmometer was based on the Leibnitz stepped wheel which had been designed at the end of the seventeenth century.[9] Multiplication and division were performed by repeated addition and subtraction respectively. The multiplicand was set on the machine using a series of slides at the front of the device. The positioning of each slide moved a pinion with ten teeth along the length of a Leibnitz stepped wheel. When the handle of the machine was turned the pinions were brought into gear with the stepped wheels. As the stepped wheels made a complete rotation the pinions turned through an angle proportional to the digit set on the slide. The movement of the pinions caused the figure wheels at the back of the machine to rotate by an equivalent amount and thus display the result of the addition or subtraction. A tens-carry mechanism then came into play. The number of handle turns made (i.e. the multiplier) was recorded by a set of figure wheels situated between the results figure wheels and the setting slides. The driving mechanism of the machine could be moved to the left to achieve multiplication in any desired position. A lever situated at the top left of the machine selected addition/multiplication or subtraction/division by reversing the driving mechanism.

The Thomas Arithmometer was used mainly by banks, shops, and other businesses for the addition of many numbers. By 1878 over 1500 Arithmometers had been manufactured and sold (Martin 1925). Following the success of Thomas's Arithmometer several machines of this type appeared on the European market which embodied many improvements while retaining the basic design. Some of the better known arithmometer type machines on sale in Britain were the Burkhardt Arithmometer (1878), the Tate (1883), the Saxonia (1895), the Peerless (1904), the Archimedes (1906), the TIM (1906), the Madas (1908) (Plate 5), and the Rheinmetall (1924).[10]

Towards the end of the nineteenth century Baldwin in the United States and Odhner in Europe patented similar machines in which the Leibnitz stepped wheel was replaced by a much smaller pin wheel. Each pin wheel represented a digit of the mutiplicand in the same way as each stepped wheel had done in the Arithmometer. The multiplicand was set on the machine by levers which caused a corresponding number of pins to project from each pin wheel. When the handle of the machine was turned the pin wheels geared with the figure wheels of the product register and advanced them through the number set on the levers.

Like Arithmometers, Odhner, or pin wheel, type machines performed multiplication and division by repeated addition and repeated subtraction respectively. Odhner type machines usually had two registers, a multiplier register and a product register, with capacities of up to 10 and 18 digits respectively. Odhner type machines had two advantages over arithmometer machines. Firstly, in Odhner type machines subtraction could be performed by simply turning the crank handle backwards and was a natural operation to perform. In arithmometers a sliding knob, or lever, had first to be placed in the subtraction position before turning the crank handle; the extra movement meant that subtraction was a slower process on arithmometers than on Odhner type machines. Secondly, Odhner type machines were smaller and lighter than arithmometers because the pin wheel mechanism was very much more compact than the Leibnitz stepped wheel.

In 1892 Grimme Natalis and Company in Germany began to manufacture Odhner type machines under the trade name 'Brunsviga' and many were exported to England (Plate 6). The Brunsviga rapidly became a popular machine and, by 1912, 20 000 machines had been produced (Baxendale 1929*a*). It remained a much used device for over 50 years. Machines of the Brunsviga type were sold under many different names including Monopol-Duplex (1894), Bernolina (1901), Triumphator (1904), Marchant (1911), Colt's Calculator (1912), Rema (1915), Facit (1918), Hannovera (1921), Britannia (1922), Eos (1923), and Mulvido (1924).[11]

In the late 1880s Eugene Felt started to manufacture the Comptometer. The Comptometer was a key-driven device; that is the depression of a key actuated the adding mechanism. The adding mechanism consisted of a series of segment levers. A key depression both actuated the mechanism and acted as a stop which determined how far the segment lever travelled during the machine cycle. The movement of the segment lever controlled, through a series of gears, the rotation of the figure wheels in the product register. Immediately following addition the tens-carry mechanism came into operation.

The Comptometer was used mainly for the rapid summation of columns of figures in accounts work. Because the Comptometer was a key-driven device, addition could be performed very quickly because the action of turning a handle in order to actuate the adding mechanism was eliminated. Many figured numbers could be totalled as quickly as those with a single digit as experienced operators were able to press all the necessary keys simultaneously. The Comptometer was, therefore, very popular with banks and accountants where the majority of the work involved the summation of numbers of many figures. The Comptometer could also be used for subtraction by keying in complements. Multiplication and division were carried out by repeated addition or subtraction.

The inconvenience of having to use complements on the machine for subtraction and division, meant that the Comptometer, and other key-driven machines which subsequently appeared on the market, were less popular with casual scientific users than the Brunsviga or arithmometer machines. In an article in *Nature* Boys (1901) concluded that 'the comptometer . . . is not so convenient as the arithmometer for reducing and computing observations in the laboratory' (p.268). As a further criticism of the machine Boys wrote 'the comptometer makes a most aggravating noise, like a typewriter through a megaphone; but other arithmometers are noisy, none, however, are as bad as this machine' (p.268).

Adding machines with printing facilities (commonly called adding and listing accounting machines) were first produced by Felt in 1889 and Burroughs in 1892. These machines quickly became popular with banks because they provided a printed record of the additions and subtractions the machine performed. They were therefore used to produce statements of account for bank customers (Turck 1921 and d'Ocagne 1928).

The Burroughs Adding and Listing Machine (Plate 7) operated using a series of levers pivoted at their centre. At one end of each lever was a set of figures 0–9 and at the other a segmental rack which could be thrown into gear with the number wheels of the machine's total register. The depression of a key limited the angle through which the associated lever could move. Once a number had been set on the keyboard of the machine a handle was turned to activate the printing and the adding mechanisms. The individual levers pivoted until stopped by the key setting. Spring-actuated hammers then struck the levers to type one of the figures 0–9. The segmental racks were then thrown into gear with the number wheels of the total register. The movement of the lever returning to its original position advanced the number wheels by the amount set on the keyboard. Once the addition had been completed a tens-carry mechanism came into operation. In this way numbers set on the keyboard were listed and simultaneously accumulated in the total register.

To read from the total register the total key on the keyboard was depressed. This caused the segmental racks to move in gear with the figure wheels of the total register until each wheel showed zero. The spring-actuated hammers would then strike against the other end of the levers and print a figure 0–9 which represented the value which had been held by the total register.

Adding and listing machines were not immediately taken up by scientific computers because, in general, they were restricted to addition and subtraction (multiplication by repeated addition being inconvenient to achieve). However, by the 1920s they began to be applied to table-making (see Chapter 3).

A slightly different type of calculating machine which became commercially available in 1899 was the Millionaire. It was based on a machine designed by L. Bollée eleven years previously (Martin 1925). The Millionaire was unusual in that it did not perform multiplication by repeated addition but used a mechanical multiplication table. Once the multiplicand, or dividend, had been set on the machine the crank handle had only to be turned once for each figure in the multiplier, or divisor. The Millionaire was therefore a very fast machine for multiplication and division. It was, however, about four times as big as the standard Brunsviga and was sometimes sold complete with legs to form a desk. The Millionaire was approximately twice as expensive as the Brunsviga (McCarthy 1924).

This brief overview illustrates that by 1900 several types of calculating machine were on the market and, as original patents expired, many different makes began to appear. Martin (1925) lists 229 different types of calculating machine commercially available during the period 1820–1925. The majority of the calculating machines on sale in Britain were manufactured in Germany or the USA with a much smaller number being made in Switzerland and Austria. Very few calculating machines were made in Britain but most of the basic designs were available under a variety of trade names.

THE INTRODUCTION OF DESK CALCULATORS TO SCIENTIFIC WORKERS

Calculating machines, particularly the Brunsviga, were thus available to scientists and engineers at the beginning of the twentieth century. For example, the Brunsviga was being advertised in *Biometrika*, a journal for the statistical study of problems arising in biology, as early as 1901. Some scientists very quickly replaced their logarithmic tables and slide rules with desk calculators but many others did not. Comrie (1925*a*, p.243) cites Boccardi's *Guide du calculateur* (1902) in which the author 'speaks very disparagingly' of calculating machines as an example of the type of criticism surrounding their use at that time. Comrie does, however, point out that Boccardi's experience of calculating machines had been limited to early and crude models. In contrast it is evident from a remark made in the journal *Observatory* that the Brunsviga, and other calculating machines, were known to astronomers by 1905. The author of 'From an Oxford Notebook',[12] which appeared regularly in *Observatory*, enthusiastically described the use of Brunsvigas in the Egyptian Survey Office in Cairo and applauded their efficiency for routine work (Anon. 1905, p.362). Although the author assumed that his readers, professional and amateur astronomers, were aware of what a Brunsviga was, the tone of the article did not suggest that all would be familiar with its use.

A few individuals did adopt calculating machines for their work during the early part of the century and subsequently introduced them into the university laboratories where they worked. One of the most important to do so was Karl Pearson, the statistician and tablemaker. In 1903 Pearson used part of a biometric research grant to buy a Tate Arithmometer and went on to use desk machines for the very substantial amount of statistical work he performed at the Biometrics Laboratory, University College, London. Pearson himself favoured the Brunsviga and photographs of Pearson as early as 1910 sometimes show a machine beside him.[13]

E. T. Whittaker, professor of mathematics at Edinburgh University, was another important and influential figure. In 1913 he set up a Mathematical Laboratory at Edinburgh and began to run lecture courses on numerical mathematics and computation.[14] The laboratory was well equipped with mathematical tables and a wide range of calculating machines including an Archimedes, several Brunsvigas, a Burroughs adding and listing machine, a Comptometer, a Mercedes-Euklid, a Millionaire, and a Tate Arithmometer. Unlike the Biometrics Laboratory at University College, London, which used calculating machinery as a tool to carry out statistical research, the Edinburgh Mathematical Laboratory was created for mathematicians in the university to use for their work. The main interest of the laboratory was numerical records in general and not fast techniques to aid practical computation. Indeed somewhat surprisingly considering the machines with which the Edinburgh Mathematical Laboratory was equipped, Whittaker, in the preface to the classic *The calculus of observations* (Whittaker and Robson 1924), recommended that mathematical tables be the main tools of those involved in computational work. He did, however, acknowledge that the use of desk calculating machines could save time and effort in some computations. Whittaker's somewhat academic experience of computation is illustrated by some of the methods given in *The calculus of observations* which were not practical where large problems were involved. The most obvious example is the method given to solve simultaneous equations using determinants (p.75).[15]

Although primarily a mathematician rather than a scientist who had to carry out calculations as a part of his work, Whittaker recognized the importance of calculating machines and gave his students the opportunity to gain familiarity with machines in the Mathematical Laboratory. This was, perhaps, the most important aspect of Whittaker's work as it was due to his influence that numerical computation courses were begun in other British Universities principally at Imperial College, London, by H. Levy and at Leeds University by G. Smeal (Rosenhead 1954*a*).

THE ADOPTION OF CALCULATING MACHINES BY SCIENTIFIC WORKERS (1913–1925)

Despite the work of Pearson and Whittaker and the introduction of calculating machines into their laboratories, there are clear indications that the use of desk calculating machines among British scientists did not become widespread until at least the mid-1920s. The 1912 report of the Royal Astronomical Society included an article by H. C. Plummer which dealt with tables of logarithms of numbers and trigonometrical functions which illustrated that logarithm tables were the main computing tool used by astronomers in the 1910s.

The activities of the British Astronomical Association (BAA) confirm that desk calculators were not common among astronomers even by the early 1920s. In 1920 the council of the BAA asked L. J. Comrie to undertake prediction calculations for phenomena of the satellites of Saturn for 1920–3 (Massey 1952, Porter 1953). This led to the proposal that a computing section be formed within the BAA to perform 'work, especially predictions, which are in some cases a little difficult for the amateur observer, and which lie outside the scope of the professional astronomer' (Comrie 1921, p.1). The section was constituted in October 1920 with Comrie as its first director supervising 24 volunteer members working primarily with tables of logarithms. The purpose of the computing section was 'to help other members of the Association by undertaking calculations beyond the powers of such applicants, or by supplying formula, or data, or information about tables' (British Astronomical Association 1921*b*, p.52). The work of the computing section was carried out by individual volunteers, most of whom were amateurs, using tables of logarithms. There was no expectation that BAA members would have access to desk calculators.

One of the most striking examples of an institution which carried out large scale computations, but which did not quickly adopt calculating machines, was the Nautical Almanac Office (NAO). The NAO annually published navigation tables and ephemerides which involved a very large amount of computation. Up to 1926 the work was, on the whole, performed using logarithm tables. There are two main reasons why the NAO did not adopt calculating machines to carry out its routine work. First, the work of the NAO consisted primarily of constructing tables by interpolation and checking the tables by forming and examining their differences. As differencing requires only simple subtraction, the process would not have benefited from the application of calculating machines, although the interpolation procedure would have been simplified. Secondly, and more significantly, most of the work carried out by the NAO was performed outside the Office by retired members of staff (Greaves 1953, p.295). Surprisingly this system was easy to administer and maintain as those involved were familiar with the work. The main disadvantage was that the geographical distribution, and possibly the age of the workers, made it very difficult to introduce new methods into the Office.

Another example of an organization which might have been expected to use modern computing techniques was the British Association for the Advancement of Science Mathematical Tables Committee (BAASMTC). The committee was set up in the 1870s (just as calculating machines were beginning to appear) to report on and, if necessary, compute mathematical tables. Although the committee published tables on a fairly regular basis in the British Association Annual Reports, they did not apply calculating machines to do their work until L. J. Comrie joined the Committee in 1928 (Anon. 1948).

The slow adoption of calculating machines by scientists in general is further illustrated by the *Handbook of the Napier tercentenary celebration* (Horsbrugh 1914). In the *Handbook* the section dealing with mathematical tables covered 30 pages half of which was devoted to a history of mathematical tables in keeping with the Napier tercentenary celebrations. The other 15 pages gave a working list of mathematical tables with the emphasis on tables 'useful to those actually engaged in computation' (p.47). In contrast, the calculating machine section covered 65 pages which included general articles on calculating machines and their capabilities plus detailed descriptions of all the principal types of machine available. No analysis was given of the suitability of different types of machines for particular types of calculation. The contrasting ways in which mathematical tables and calculating machines were treated in the *Handbook* beg the conclusion that in 1914 calculating machines were still a novelty and mathematical tables a commonplace tool.

There were four principal reasons why desk calculating machines were not widely used by scientists before the mid-1920s. Firstly the machines were, in the early 1900s, quite costly in comparison with books of tables. In 1901, an 18 × 10 figure Brunsviga cost £27.10*s*.0*d*. and a Tate Arithmometer £40. Mathematical tables could be purchased for anything from 4*s*. to £2. When it is considered that the average income per capita in 1900 was approximately £40 per annum (Mathias 1983), a desk calculator was relatively expensive. The high cost of a desk calculating machine was acceptable in a mathematical or statistical laboratory where machines could be made available to a number of workers and could, therefore, be almost continually in use, but desk machines were expensive for the individual. Thus Pearson and Whittaker could make use of machines while the mathematicians of the BAASMTC or astronomers of the Royal Astronomical Society and British Astronomical Association, who worked as individuals, could not.

Secondly, many of the first machines to appear on the market were large and cumbersome to use (they often took up the complete working surface of a desk and needed two men to lift) and some of the very early models were also unreliable or difficult to use. As designs improved, however, the machines became very dependable (Comrie 1925*a*). They could also be regularly serviced by the manufacturer or selling agent thus increasing their reliability. Improved manufacturing and engineering techniques also led to a reduction in the size and weight of the machine. M. R. Williams (1982) notes that the compactness and light weight of a machine became a manufacturer's selling point and this is confirmed by the data presented in McCarthy's catalogue of business machines published in 1924.

Thirdly, calculations performed using desk calculating machines required natural and not logarithmic values of the trigonometrical, exponential, and hyperbolic functions. Because the bulk of all computation done since the time of Napier had been performed using logarithms there was a considerable range of logarithm tables readily available. By 1918, in a response to the increasing need for tables of the natural values of trigonometrical functions, at least five sets of such tables had been published.[16] Of particular interest is Lohse's *Tafeln fur Numerisches Rechnen mit Mashinen* (1909) which were constructed especially for use with desk calculating machines. Gifford's tables of *Natural sines* (1914) were also published to provide tables which could be used with desk calculators. In 1915 Gifford summed up the situation regarding the need for natural tables. She wrote 'I suppose the general use of logarithms had to wait until logarithmic tables of sines and tangents were compiled, just as now the use of calculating machines is hampered by the want of natural numbers (E. Gifford, 'New tables of natural sines', in Knott 1915 pp.287–98).

Fourthly, while calculating methods for use with logarithms were well established those using desk machines were not. This was obviously an obstacle to the adoption of calculating machines. Comrie later, in the 1920s and 1930s (see Chapter 3), developed many techniques for use with desk calculating machines which increased their usefulness. Pearson and the staff of the Department of Applied Statistics, University College, London went some way towards making other scientists aware of practical numerical methods suitable for use with desk calculating machines. In 1919 the Department of Applied Statistics started to produce a series of booklets under the title *Tracts for computers* which were intended to be short monographs on various mathematical techniques including interpolation, quadrature and mechanical integration, calculating machines, bibliographies, and new or reprinted tables designed to be of practical help to those engaged in computation on any scale (K. Pearson 1919). However, only 5 of the 26 *Tracts* in the series described mathematical techniques. Numbers 2 and 3 dealt with the subject of tablemaking and interpolation. *Tracts* 6, 7 and 8 discussed smoothing, the evaluation of the incomplete β-function, and quadrature and cubature respectively. *Tract* 13 gave a bibliography of mathematical tables. The remaining *Tracts* consisted of a variety of mathematical tables. Of these, *Tract* 5 was the most relevant to those performing computations as it gave a table of coefficients for Everett's central difference interpolation formula.

As desk calculating machines improved and suitable natural value tables of common functions became available, the use of desk machines by scientific workers spread. For example, during the First World War A. V. Hill led a team, which included D. R. Hartree, which applied desk machines to anti-aircraft gunnery research. The Department of Applied Statistics at University College, London, which already used calculating machines in its own statistical work, offered its services to the Government on the outbreak of war and performed calculations for the Department of Trade, the Royal Aircraft Factory, Farnborough, the Admiralty Air Department, the Anti-Aircraft Section of HMS *Excellent* (A. V. Hill's group), the Ordnance Committee, and the Ministry of Munitions (K. Pearson 1918).

By the mid-1920s the use of calculating machines had become more common as scientists became more aware of their capabilities. In his 1925 report on mathematical tables to the Royal Astronomical Society Comrie noted

> the appearance of a remarkable group of tables, thoroughly indicative of modern tendencies. They indicate very clearly that progress is being made . . . in the adoption of calculating machines by scientific computers, which is made manifest by the publishing of extensive tables of the natural values of the trigonometrical functions. (Comrie 1925*b*, p.386)

Seven years later Comrie felt able to report stronger evidence of the use of desk calculating machines to the Royal Astronomical Society. He stated that recently published tables

> reflect the modern tendency towards the use of calculating machines in three ways – first, by the extensive use of machines in their making, as reported in prefaces; secondly, by the fact that logarithmic values are tending to be superseded by natural; and thirdly, because the possession of a machine is virtually assumed in tables of special functions, where a wide interval is used and several even orders of differences provided. (Comrie 1932*b*, pp.338–9)

It is fair to conclude that by the early 1930s desk machines had replaced 4-, 5-, 6-, and 7-figure logarithms as the most common method of large scale computation used both by individuals and institutions, with the acceptance of the use of mechanised computing methods came the gradual organization and centralization of computing resources.

ENDNOTES FOR CHAPTER 2

1 Whittaker and Robson 1924.

2 References given under Barlow 1814.

3 Logarithms are divided into two parts; the characteristic and the mantissa. For example, in $\log_{10} 2 = 0.3010300$, 0 is the characteristic and .3010300 is the mantissa. In logarithms to base 10 the value of the mantissa is independent of the position of the decimal point in the original number; the position of the decimal point is determined by the characteristic. This is not the case with logarithms to base e or base e^{-1}.

4 Vega's *Thesaurus logarithmorum completus* . . . also incorporated a new edition of Vlacq 1628.

5 For a history of the slide rule see Cajori 1909.

6 By, for example, Cajori 1909; Stokes 1914; and Goldstine 1972.

7 In mathematical tables, desk machines, and, today, computers numbers are represented as individual digits. Arithmetic is therefore performed to an accuracy determined by the number of digits the table or machine can record. Today these devices are described as 'digital'.

8 Until 1957 when Franz Hammer discovered the prior claim of William Schickard's calculating machine of 1623, Pascal's machine had been regarded as the first practical invention of a calculating machine. Unlike Schickard's machine, however, Pascal's calculator had a significant influence on the subsequent development of calculating machinery (Goldstine 1972).

9 The Leibnitz stepped wheel consisted basically of a cylinder. Cogs of different lengths ran along the surface of the cylinder parallel to its axis. The cogs represented the values 0–9. A cogged wheel could be set at different distances along the cylinder so that by bringing the cylinder and the wheel into gear with each other a complete revolution of the cylinder would cause the wheel to turn by the required number of positions 0, 1, 2, . . . , or 9.

10 The main sources of information regarding calculating machines are Turck 1921; McCarthy 1924; Martin 1925; d'Ocagne 1928; and Baxendale 1929*a*.

11 Turck 1921; McCarthy 1924; Martin 1925; d'Ocagne 1928; and Baxendale 1929*a*.

12 'From an Oxford Notebook' was written anonymously by H. Turner, Savilian Professor of Astronomy at Oxford University 1893–1930, and was an amusing account of Turner's activities, travels, astronomical observations, and comments on life.

13 Yule and Filon 1938, facing p.70, and E. S. Pearson 1936, facing p.222.

14 The term 'numerical analysis' did not come into use until after the Second World War.

15 In *Numerical analysis* (2nd edn 1958, p. 166) Hartree noted that the direct evaluation of determinants is 'never' the best way to solve a set of simultaneous equations.

16 Andoyer, H. *Nouvelles tables trigonometriques fondamentales (valeurs naturalles)*, 3 vols, 1915, 1916, 1918. Hermann, Paris. Burrau, C. *Tafeln der Functionen Cosinus und Sinus*. Reimer, Berlin. 1907. Gifford, E. *Natural sines*. Manchester 1914. Lohse, O. W. *Taflen fur numerisches Rechen mit Machinen*. Engelmann, Leipzig. 1909. Peters, J. *Einundzwanzigstellig Wert der Funkionen Sinus und Cosinus*. Reimer, Berlin. 1911.

3

The mechanization of computation at the Nautical Almanac Office

During the 1910s and 1920s scientists were, in general, slow to adopt desk calculating machines for their work. Small computing laboratories had been set up by Pearson and Whittaker but in most other institutions routine computations were carried out using logarithmic tables. As outlined in the previous chapter the British Nautical Almanac Office (NAO) at Greenwich was no exception.

Much of the work carried out by the NAO involved the routine tabulation of annually published ephemerides. The work was shared internationally by five offices based in Britain, America, Germany, France, and Spain.[1] The British Office had the task of providing the predicted positions of the Sun, Moon, and planets and the apparent places of the stars. For some tables all the values tabulated were individually computed but, in order to reduce the amount of computation which had to be performed, certain tables were produced by interpolation between independently calculated pivotal values.

In the 1920s most of the computational work carried out by the NAO was performed by hand using logarithm tables. Desk calculating machines were not used. In fact much of the computational and proof-reading work was performed outside the Office itself by retired members of staff – a system which had been introduced by P. H. Cowell, Superintendent of the NAO 1910–30. One of the few exceptions to the almost complete lack of mechanized computing methods in use at the NAO during the 1910s was a Burroughs Adding and Listing machine installed in 1912 by T. C. Hudson. This machine was specially commissioned by the NAO to make it suitable for their work and could perform addition and subtraction (then a novel feature) in either decimal or sexagesimal notation. Hudson applied the machine to the subtabulation of functions from previously computed pivotal values. Hudson (1914) described the use of the Burroughs machine for building up a table of the daily Heliocentric places of Mars where every twelfth table entry had been independently computed.

Consider the table for 1923. The values for 13 January and 25 January are 31° 21′ 51″.31 and 38° 20′ 18″.53 respectively. Hudson derived a graphical method of interpolation which gave the last digits of the intervening values (shown in Appendix 1.1). From the last digits of the function values, the last digits of the first and second differences could be easily deduced (Appendix 1.2). From that part of the table already computed the first and second differences above 13 January were known to be $-35' \, 13''.51$ and $-3''.22$ respectively. From this information the complete second difference could be inferred (Appendix 1.3).

By setting the second differences on the Burroughs machine the first differences could be accumulated using the relation $D^{I}_{1/2} = D^{I}_{-1/2} + D^{II}_{0}$ where D^{I} = first differences and D^{II} = second differences. The machine printed results in the form shown by Appendix 1.4. The function values, in this case the Heliocentric longitude of Mars, could not be calculated simultaneously because the Burroughs Adding and Listing machine of the 1920s had only one accumulating register. Therefore to complete the tabulation the paper on which the first differences had been printed was reinserted into the paper carriage of the machine and the now known first differences were keyed into the machine in order to accumulate the function values using the relation.

$$f_1 = f_0 + D_{1/2}$$

As a check against transcription errors being made by the operator, the machine was set to print the first differences keyed into the machine over those originally calculated (Appendix 1.5). In this way any transcription errors could be easily spotted. A second check on correctness was the reproduction of the already known value for 25 January.

The Burroughs Adding and Listing machine, which was still in use at the NAO during the late 1920s, was used only to build up tables of functions for which the full second difference could be deduced from its last digit. There were two reasons for this. First, Hudson's graphical method for finding the last digits of the interpolates was not suitable for deriving differences columns higher than the second. Secondly, the paper carriage of the machine was not wide enough to accommodate printing the function value and difference columns higher than the second across the page. This second difficulty was perhaps the most limiting in practical terms. Nevertheless the Burroughs machine was regularly used at the NAO to prepare a wide range of tables including the daily heliocentric places of Venus, the daily heliocentric place of Mars, the Moon's hourly place, and the co-ordinates of the Sun for every noon and midnight. Although this machine was used successfully it was the exception at the NAO and not the rule.

THE MECHANIZATION OF COMPUTATION AT THE NAO

In 1925 L. J. Comrie (1893–1950) joined the NAO where he was quickly appointed Deputy Superintendent and later (1930) Superintendent. During his 11 years at the NAO Comrie completely revolutionized the computing methods used at the Office. He installed desk calculators, carriage controlled adding and listing machines, and punched card machines. But, more importantly, Comrie developed efficient numerical methods for use with the mechanical computing aids. By publishing the results of his work, lecturing widely, and taking on consultancy work Comrie greatly influenced the way in which scientific computation was performed not only at the NAO but throughout Britain.

Before taking up his appointment at the NAO Comrie had already had some experience of desk calculator work. On Armistice Day (11 November) 1918 Comrie received his first lesson on a Brunsviga calculator from Karl Pearson of the Biometric Laboratory, University College, London. Pearson was a statistician and had much experience of numerical computation. He greatly influenced Comrie's opinions on the importance of maintaining high standards in computational work. In 1944 Comrie partly acknowledged his debt to Pearson by quoting from the preface of *Tracts for computers no. 1* in which Pearson's views on the necessity of providing adequate training in numerical computation were plainly expressed (Pearson 1919, pp.1–2, Comrie 1944). It is clear from his work over the following 30 years that Comrie adopted the standards set by Pearson.

After serving with the New Zealand Expeditionary Forces during the First World War (in which he lost a leg), Comrie enrolled as a postgraduate student in astronomy at Cambridge in 1919. He had previously graduated from University College, Auckland, New Zealand in Chemistry in 1915. While a student at Cambridge Comrie spent two months working at the Royal Observatory at Greenwich. Greaves, Comrie's obituarist, brother-in-law, fellow student, and Astronomer Royal for Scotland (1938–50), reported that, even at this early stage of his career, Comrie was forthright in condemning the computing methods used at Greenwich (Greaves 1953, p.297). Comrie also took an active role in the work of the British Astronomical Association (BAA). In 1920 he set up the computing section of the Association and organized computations carried out by 24 volunteer members (see Chapter 2). While still a postgraduate student Comrie had, therefore, gained considerable experience in organizing and administering large computations. In 1922 Comrie resigned his directorship of the BAA computing section and was succeeded by Major A. E. Levin.

Comrie received his Ph.D. in astronomy in 1923 and left Cambridge to spent two years teaching astronomy in the United States; first at Strathmore College, Philadelphia, and later at North Western University, Illinois. At both institutions he introduced practical computing courses into the curriculum. Comrie also spent some time studying the advantages of mechanical computation and, during his time in the United States, he published the first of many papers on the use of desk machines for scientific computation (Comrie 1925*a*).

Comrie returned to England in 1925 and took up a post as an assistant at the NAO in October of that year. In the following March Comrie was appointed Deputy Superintendent. As Cowell, the Superintendent at the NAO, was due to retire in 1930, Comrie was practically assured of succeeding him. D. H. Sadler, Comrie's successor as Superintendent of the NAO, recalled that very soon after his appointment as Deputy Superintendent, Comrie began to make his presence felt (Greaves 1953, p.297). Almost immediately Comrie began to install calculating machines at the NAO, to develop numerical techniques for their application to the work of the NAO, to build up a staff of machine operators, and to gradually change the methods of computation used.[2] By June 1927 an electrically powered Monroe, a Nova-Brunsviga, and a Comptometer had been installed at the NAO. At first these machines were used simply to speed up processes previously done mentally or performed using logarithmic tables. But replacing tables in this way was not sufficient. Comrie began to develop numerical methods which reduced much of the work performed at the NAO to a series of routine steps which could be carried out by junior staff. Even where suitable desk or accounting machines were not available Comrie revised existing manual computing methods to make them more efficient.

As Deputy Superintendent of the NAO during 1926–30, Comrie's work in revising the computational methods used at the NAO took on three important aspects. First, he developed the end-figure method of interpolation. Second, he applied the Brunsviga-Dupla and the Burroughs Class 11 Adding and Listing machine to subtabulation from second differences. And third, Comrie installed Hollerith punched card machines for the computation of tables of the position of the moon.

COMRIE'S END-FIGURE METHOD OF INTERPOLATION

Because interpolation was such a fundamental part of the work of the NAO, it was important that the process be performed as efficiently as possible. In the 1920s the two interpolation techniques most commonly used were the Lagrangian method and the building up of interpolates by continuous summation from columns of finite differences derived from the Everett, the Newton-Bessel, the Newton-Stirling, or the Newton-Gauss interpolation formulae. Comrie objected to both procedures on the grounds of laboriousness and poor accuracy. Hudson's method of interpolation using a Burroughs Adding and Listing machine was in use at the NAO at that time. In Comrie's view there were two disadvantages to Hudson's method. First, the graphical way in which Hudson derived the last figure of the interpolates required some degree of expertise and experience. It could not be carried out routinely by junior staff. Secondly, to tabulate a function from its second differences the Burroughs machine had to be used twice. Being fitted with only one register it did not have the capacity to accumulate the first differences and the function value simultaneously. The method was also limited to functions where the third and higher differences were not significant.

These difficulties led Comrie to develop a method of interpolation which could be carried out routinely for a wide range of functions. Comrie's method was based on Bessel's interpolation formula and it was applicable to interpolation to tenths and fifths. It was inspired by Hudson's earlier method but instead of finding the last figure of the interpolates graphically, Comrie developed a way of numerically deriving the last two figures of each interpolate. Comrie called this process the end-figure method of interpolation.

Interpolation by Comrie's end-figure method was quite straightforward to carry out. Comrie published the method in 1928 in a paper to the Royal Astronomical Society (Comrie 1928*b*) and again as an appendix to the 1931 *Nautical almanac* (Comrie 1929*a*) where he also presented the tables of end-figures necessary to carry out the process. Essentially the end-figure method of interpolation consisted of eight steps.

1. Difference the original table of pivotal values.
2. Use the two end-figures of these differences as arguments to look up the end-figures of the constituent terms of the Bessel interpolation formula in previously computed interpolation tables.
3. Sum the end-figures of the Bessel interpolation formula terms to give the end-figures of the interpolates.
4. Difference the end-figures of the interpolates to get the end-figures of the first, second, third, and fourth differences. Test the fourth differences for smoothness to check for errors.
5. From the end-figures infer the whole of the second differences.
6. Calculate the leading first difference using the formula $0.1D^{\mathrm{I}}_{1/2} - 0.0225 (D^{\mathrm{II}}_0 + D^{\mathrm{II}}_1) + 0.006D^{\mathrm{III}}_{1\frac{1}{2}} + 0.004(D^{\mathrm{IV}}_0 + D^{\mathrm{IV}}_1)$.
7. From the leading first difference and the inferred second differences find the full first differences by continuous summation.
8. Find the full values of the interpolates by continuous summation from the first differences.

Comrie's end-figure method was an important part of his revision of computing methods at the NAO because it provided an accurate and clearly defined way of interpolating tables. There was, however, no advantage in using desk calculators for the derivation of the end-figures as only trivial two figure additions were required. Differencing the end-figures required only simple subtractions and so, again, little was to be gained by the use of desk calculators. The only part of the process which could be mechanized successfully was the final stage of the procedure, that of building up the function by continuous summation from the derived differences. Mechanizing the sub-tabulation of a function from its second differences was the second major aspect of Comrie's work as Deputy Superintendent.

SUB-TABULATION OF A FUNCTION FROM ITS SECOND DIFFERENCES

Hudson's Burroughs machine, already installed at the NAO, could have been used for sub-tabulation using differences derived by Comrie's end-figure method. However, the machine had to be used twice for each function tabulated and both the first and second differences had to be manually entered into the machine. This made the process very time-consuming and Comrie searched for a machine which would produce the required table without the computed first differences having to be reset on the machine.

By March 1928 Comrie had developed a method of building up a function from its second differences using a Brunsviga-Dupla calculating machine which first appeared on the market in 1927. Like the other Brunsviga machines on the market the Brunsviga-Dupla possessed setting levers, a multiplier register and a product register R_1. A turn on the handle of the machine would increment the multiplier register and add the contents of the setting levers into the product register R_1. The contents of the product register could be directly transferred to the setting levers. In addition the Brunsviga-Dupla had two important features which made it particularly suitable to the building up of a function by continuous summation of its differences and which did not appear together in other calculating machines of that period. First, it had a second product register, R_2. When the handle of the machine was turned the number on the setting levers was added simultaneously into both product registers (it was possible to suppress addition into R_2). Secondly, it was possible to set a number directly into the first product register, R_1, via a set of thumb wheels without having to disturb the setting levers. It was the combination of these features and the facility which allowed the contents of the product registers to be transferred to the setting levers which made the Brunsviga-Dupla suitable for the building up of a function from its second differences. The method Comrie used was quite straightforward.

Given

$$f_1 = f_0 + D^{I}_{1/2}$$

and

$$D^{I}_{1\frac{1}{2}} = D^{I}_{1/2} + D^{II}_{1}$$

where f = the function value

D^{I} = the first differences

and D^{II} = the second differences.

Then if

the multiplier register held the argument,
the product register R_2 held the function,
the setting levers held the first difference,
and the product register R_2 held the second difference,

the procedure was as follows:

1. Set the original function value, f_0, onto the setting levers and turn the handle. This set f_0 into the product registers R_1 and R_2 and incremented the multiplier register to show 0.
2. Clear R_1.
3. Set the first difference, $D^{I}_{1/2}$, onto the setting levers.
4. Set the second difference, D^{II}_{1}, into the first product register, R_1, via the thumb wheels.
5. Turn the handle. This causes
   (*a*) the first difference, $D^{I}_{1/2}$, to be added to the function value, f_0, in R_2 to form the new function value f_1;
   (*b*) the first difference, $D^{I}_{1/2}$, to be added to the second difference, D^{II}_{1}, in R_2 to form the new first difference $D^{I}_{1\frac{1}{2}}$;
   (*c*) the argument in the multiplier register to be incremented by 1.
6. Read the new function value and first difference from the machine.
7. Transfer the new first difference $D^{I}_{1\frac{1}{2}}$ from R_1 to the setting levers and clear R_1.
8. Cycle from step 4.

By using complements and the machines facility to show negative values in red, the procedure could be maintained in cases where the first and/or second differences were negative. The main disadvantage of using the Brunsviga-Dupla to build up tables was that the machine lacked printing facilities. Copying down the results from the machine by hand introduced the risk of transcription errors occurring in the completed table.

In 1929 Comrie installed a Burroughs Class 11 Adding and Listing Machine at the NAO and immediately began to use the machine for the preparation of tables for publication. Unlike the early Burroughs machine, which had only one accumulating register, the Class 11 had both a register and a crossfooting device.[3] The crossfooter was the principal accumulating mechanism of the machine and numbers from the keyboard could be either added into or subtracted from it. The register was also an accumulating device but could perform addition only. The contents of both the crossfooter and the register could be printed using the total and subtotal keys on the machine.[4] The number held in the crossfooter could also be added into the register and vice versa.

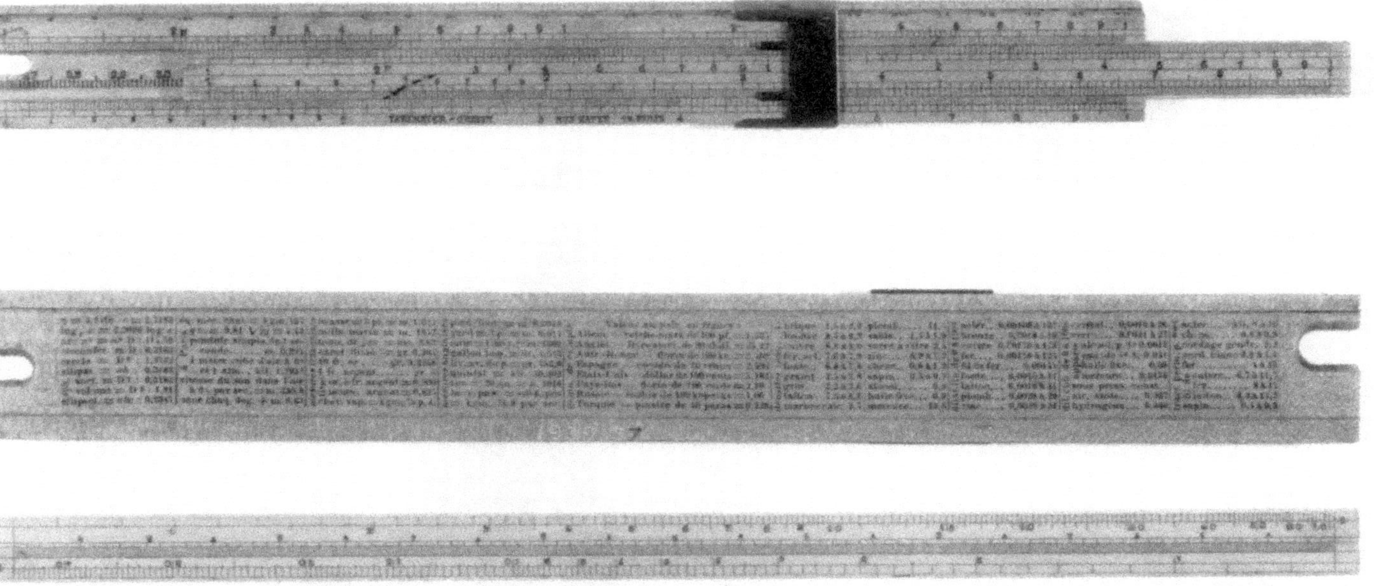

Plate 1. Mannheim slide rule, front and back views. Reproduced by permission of the Trustees of the Science Museum, South Kensington. Negative no. 374/60.

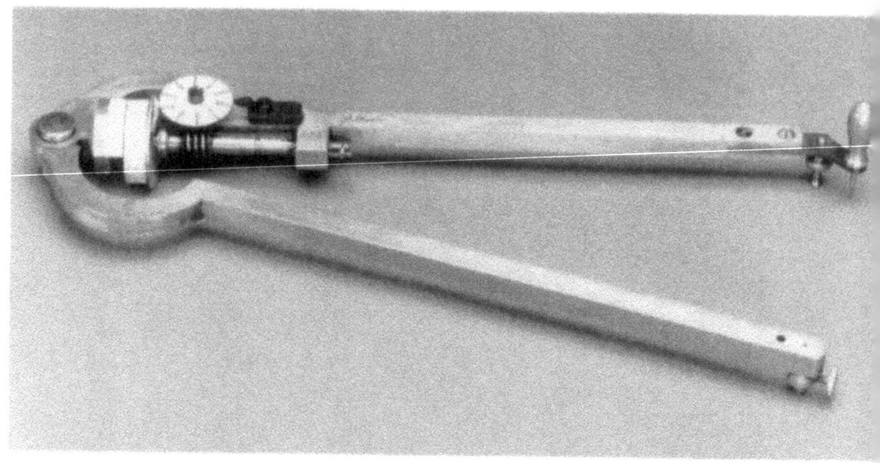

Plate 2. Amsler Polar Planimeter. Smithsonian Institution. Photo no. 85-9 1821.

Plate 3. Coradi Integraph. Smithsonian Institution. Photo no. 79-1995

Plate 4. Kelvin's Harmonic Analyser. Reproduced by permission of the Trustees of the Science Museum, South Kensington. Negative no. 2619.

Plate 5. Madas Calculating Machine illustrating the arithometer type of machine. Smithsonian Institution. Photo no. 88-15550.

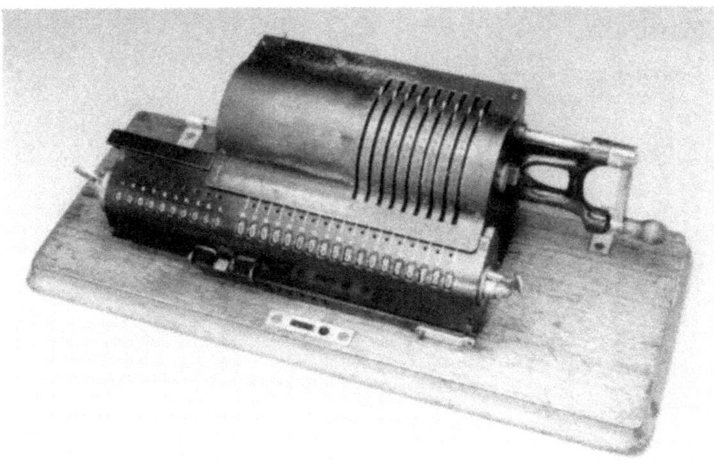

Plate 6. Brunsviga Calculating Machine. Smithsonian Institution. Photo no. 84-18558-5.

Plate 7. Burroughs Adding and listing Machine. Reproduced by permission of the Trustees of the Science Museum, South Kensington. Negative no. 3255.

Plate 8. Hollerith Tabulating Machine of the 1930s. Reproduced by permission of the Trustees of the Science Museum, South Kensington. Negative no. 915/71

Plate 9. Differential Analyser at Manchester University. Reproduced by permission of the Trustees of the Science Museum, South Kensington. Negative no. 7847.

Plate 10. Pilot Ace Computer at the National Physical Laboratory. Reproduced by permission of the Trustees of the Science Museum, South Kensington. Negative no. 596169.

Plate 11. EDSAC 1 at Cambridge University. Reproduced by permission of the Trustees of the Science Museum, South Kensington. Negative no. 289/77.

Plate 12. Mark I Computer at Manchester University. Reproduced by permission of the Trustees of the Science Museum, South Kensington. Negative no. 1350/80.

In addition to the control keys on the keyboard much of the operation of the Burroughs Class 11 was controlled by interchangeable stop rods and control rods attached to the paper carriage of the machine. The stop rods were used to divide the paper carriage into several tabular positions as well as control the carriage return. The control rods were used to activate levers which caused the crossfooter and the register to add in any particular carriage position. This allowed cycles of operations to take place across the carriage of the machine. Comrie used this feature in applying the Burroughs Class 11 machine to the building up of functions from their second differences.

Comrie set up the carriage of the Burroughs Class 11 into three positions.

Position 1: Crossfooter – non-add;
Register – add;
Print columns 1–17.
Position 2: Crossfooter – add;
Register – non-add;
Print columns 1–13.
Position 3: Crossfooter – add;
Register – add;
Print columns 1–13;
Carriage return with linefeed.

For the tabulation of a function from its second differences the register held the function values and the crossfooter the first differences. The table itself was printed out in a non-standard way. The usual tabulation would have printed

$x \quad f \quad D^{\text{I}} \quad D^{\text{II}}$

across the page where

$x =$ the argument
$f =$ the function value
$D^{\text{I}} =$ the first differences
and $D^{\text{II}} =$ the second differences.

Comrie adopted the layout

$x \quad f \quad D^{\text{II}} \quad D^{\text{I}}$

in order to simplify the machine operation.

The tabulation was achieved relatively simply.

1. Carriage position 1. Set the initial argument on the keyboard using columns 14–17 (the left-most columns). Set the initial function value on the remaining columns 1–13. Actuate the motor bar to print both the argument and the function value and add the original function value into the register.

2. By hand move the carriage to position 3. Set the first difference on the keyboard using positions 1–3. Actuate the motor bar to print the first difference and add it into both the register and the crossfooter. The register now contains the next function value and the crossfooter holds the first difference.
3. Press the return key. The carriage is now in position 1. By hand increment the argument.
4. Carriage position 1. Sub-total the register to print the argument and the new function value. The carriage tabulates.
5. Carriage position 2. Increment argument by hand. Set the second difference using columns 1–13. Actuate the motor bar to print the second difference and add it into the crossfooter to form the next first difference. The carriage tabulates.
6. Carriage position 3. Sub-total the crossfooter to print the first difference and add it into the register to form the next function value. The carriage returns automatically to position 1 with a linefeed.
7. The cycle continues from step 4.

By developing this method for building up a function from its second differences Comrie mechanized and simplified a major part of the NAO's work.

THE USE OF PUNCHED CARD MACHINES TO CALCULATE THE POSITIONS OF THE MOON

The third main feature of Comrie's work while Deputy Superintendent of the NAO was the complete revision of the methods used to compute tables of the positions of the Moon. Tables giving the position of the Moon for every noon and midnight were first included in the *Nautical almanac* in 1923. The positions of the Moon were derived from the summation of a number of harmonic terms given in Brown's *Tables of the Moon* (1919). Although the existence of Brown's *Tables* made the calculation conceptually easy to perform, it was a major undertaking for the NAO keeping two members of staff fully occupied all year round. Brown's *Tables* were made up of 108 individual tables covering 660 pages and represented the harmonic terms required for specific periods. The *Tables* were arranged in order of increasing time.

The extraction of the terms required to calculate the position of the Moon from Brown's *Tables* was not difficult but it was extremely laborious, with 8 to 12 terms, each from a different table, being required for the calculation of a single value. Once the table entries for the first date in a sequence of calculations had been determined, they were copied down and totalled. The entries for consecutive dates followed serially in the tables, hence obtaining individual terms presented no difficulty. However the copying down and summation of the terms was a tedious task and one which introduced the danger of transcription errors. Each table entry was used on several occasions but identical combinations of entries did not recur. Occasionally discontinuities had to be introduced into the table reading to compensate for the inaccuracies of the tabulated periods.

In 1929 Comrie automated this process of Fourier synthesis using Hollerith punched card machines.[5] (Plate 8.) The tabular values given in Brown's *Tables* were punched onto 45 column cards.[6] A group of adjacent columns was called a field and represented a particular entity, for example date. Each card carried data identifying which table entry it represented as well as the table entry itself. Because British tabulators could not perform direct subtraction from the card, as the later American IBM machines did, all table entries were punched in both direct and complementary form to facilitate subtraction.

Once the cards had been punched it was important to check the accuracy of the transcription. In commercial punched card installations this was done using a device called a verifier which was used to 'repunch' the cards from the original data to catch any discrepancy in the punching. Comrie, however, did not use this method of verification. In order to relieve junior punch operators of the responsibility of verifying the accuracy of the punching, Comrie listed the cards on the tabulator and used experienced proof-readers already employed at the NAO to check the printed copy against the original tables. The punched cards representing any one of the tables in Brown's *Tables* were stacked individually and their original sequence retained.

The tabulator was the principal machine in the punched card system. The tabulator could be set up to read selected values from any area of the card, print the values, and/or add them into a combination of the five 9-figure registers. The tabulator performed the same function on each card which passed through the machine. It was also possible to set up the tabulator to total or subtotal selected registers when all cards referring to a particular date, for example, had passed through the machine.

Once the cards had been punched and the tabulator wired up the procedure to calculate the positions of the Moon was as follows.

1. Group the cards representing the necessary harmonic terms together by hand.
2. Stack groups of cards relating to sequential dates together and place in the tabulator. Make any necessary allowances for discontinuities in the data.
3. Pass the cards through the tabulator to sum the harmonic terms relating to each date. Print the result.
4. At the end of the run pass the cards through a machine called a sorter to separate the cards out into the individual tables. Return the cards to the table stacks preserving the original ordering.

Comrie's application of Hollerith punched card machines to the calculation of the position of the Moon was significant for two reasons. First, the application itself greatly reduced the effort and cost of annually producing tables of the position of the Moon for publication in the *Nautical almanac*. The work was begun in late 1928 when Comrie took on extra staff to prepare the cards. Transferring Brown's *Tables* on to punched cards took 6 months. The rest of the work was completed in 7 months during 1929 on punched card machines hired from the British Tabulating Machine Company (BTM) and installed at the NAO. To have completed the calculation for the following 10 year period (1935–45) was all that was strictly necessary at the time. However, as the most expensive part of the calculation had been the organization, card punching, and staff training, Comrie decided to take the calculation on to the year 2000 as an alternative to repeating the work 10 years later. Comrie further justified the continuation of the calculation by expressing the opinion that there was 'little likelihood of Brown's *Tables* being superseded before the end of the century; any acquisition to our knowledge during the next seven decades is almost certain to be expressed in the form of corrections to Brown's *Tables*, not in the form of new tables' (Comrie 1932*c*, p.706).[7] Due to the expiry of the lease on the Hollerith machines the entire tables of the Moon up to the year 2000 were not completed. Although all of the necessary cards had been prepared, the calculations for the Moon's latitude, longitude, and horizontal parallax were only taken up to 1950. The cards were stored in anticipation of the calculation being completed in the future. Comrie estimated that the cost of the computation, including the hire of the machines, came to less than £1500. The cost of manualling carrying out the work up to the year 2000 would have totalled over £6000. Therefore a great financial saving had been made (Comrie 1932*c*).

The second and, from an historical point of view, more important consequence of Comrie's application of Hollerith punched card machines was that it opened up the possibility of using this type of equipment for other types of scientific problems where the calculations were of a suitable nature and required on a large enough scale. Having demonstrated the use of punched card machines for Fourier synthesis in calculating the positions of the moon, Comrie went on to explore further uses of the machines. For example he proposed using punched cards to produce sums of products used in multiple correlations a technique which later became widely used by statisticians.

By 1930 then, Comrie had made substantial revisions to the computing methods used in the NAO, particularly in the development of the end-figure method of interpolation, his application of the Burroughs Class 11 adding and listing machine to the sub-tabulation of a function from its second differences, and his application of Hollerith punched card machines to calculating the positions of the Moon.

THE NAO AS A MECHANIZED COMPUTING CENTRE 1930–1936

On Cowell's retirement in August 1930 Comrie was appointed Superintendent of the NAO and, according to Sadler, then Deputy Superintendent, Comrie 'intensified his campaign for the increased use of modern calculating machines and methods of computation' (Greaves 1953, p.298). Comrie was an impatient, uncompromising man and had difficulty dealing tactfully with Admiralty administration. When Comrie presented proposals for change at the NAO the Admiralty, understandably, wanted to consider them at some length. Comrie saw those slow deliberations as incompetence on the part of the Admiralty and found it impossible to wait patiently for their approval. The Admiralty needed to be persuaded that change was necessary as all had appeared well under Cowell's administration and they would not have been encouraged by Comrie's attitude towards them. These problems with the Admiralty were, mainly, the result of Comrie's strong personality and his 'fanaticism' (Greaves 1953, p.294) about his work. They caused considerable friction between Comrie and the Admiralty throughout Comrie's Superintendentship.

Nevertheless, Comrie continued to mechanize the routine work of the NAO and by 1932 had replaced many of the Office's logarithmic tables with desk calculators. In addition he had built up a new staff consisting of both junior machine operators, experienced computing staff and astronomers. Comrie also continued to develop new numerical techniques for use with machines. The Burroughs Class 11 Adding and Listing machine which Comrie had installed at the NAO in 1929 could only be used for functions where the second difference column was continuous, oscillatory, or could be deduced. Because the machine had only two accumulating registers it could not be used to built up a function from its third or higher differences. For many functions the third or higher differences could be found more easily than the second. By the early 1930s larger Burroughs machines had become available, for example, the Class 16 which was equipped with six registers and a crossfooter. These larger Burroughs machines were not, however, suitable for building up a function from its sixth differences because all inter-register transfers had to be made via the crossfooter thus disproportionately increasing the number of operations needed to compute any one function value.

In 1931 Comrie overcame this problem with the National Accounting Machine Class 3000 (then sold under the trade name of Ellis) and had a machine installed at the NAO by 1933. The National machine had six registers into which numbers could be directly added from the keyboard; registers 1 and 3 accepted numbers positively or negatively. Any register could be totalled or sub-totalled and the contents of any register could be added directly into any other register. There was no intermediate crossfooting device through which transferred numbers had to pass and hence the National Accounting Machine could be efficiently applied to building up a function from its sixth differences. The destination of a number (i.e. the register into which it was to be added or subtracted), was controlled by interchangeable stops which acted on control arms in a similar manner to those on the Burroughs Class 11. The number to be printed and/or entered was supplied by the 11-column keyboard or by the selection of a register via the control keys on the extreme left of the keyboard. The machine was activated by the depression of the motor bar situated to the right of the keyboard.

To apply the National Accounting Machine to the sub-tabulation of a function from its known sixth differences Comrie used the following difference relations:

$$
\begin{aligned}
D^{V}_{1/2} &= D^{V}_{-1/2} + D^{VI}_{0}\\
D^{IV}_{1} &= D^{IV}_{0} + D^{V}_{1/2}\\
D^{III}_{1\frac{1}{2}} &= D^{III}_{1/2} + D^{IV}_{1}\\
D^{II}_{2} &= D^{II}_{1} + D^{III}_{1\frac{1}{2}}\\
D^{I}_{2\frac{1}{2}} &= D^{I}_{1\frac{1}{2}} + D^{II}_{2}\\
f_3 &= f_2 + D^{I}_{2\frac{1}{2}}
\end{aligned}
$$

where f = the function value
and D = the differences.

The final table weas printed across the page

$x \quad D^{VI} \quad D^{V} \quad D^{IV} \quad D^{III} \quad D^{I} \quad f$

In setting up the National machine for sub-tabulation of a function from its sixth differences Comrie divided the carriage into nine positions. Each position was set up, using stops, to carry out a specific step in the sequence of operations needed to carry out the tabulation. After the initial values had been set into the machine the sequence cycled, using the carriage return, until the complete table had been printed. The motor bar was used by the machine operator to tabulate to each of the carriage positions in turn. The sequence of operations was as follows:

Carriage position 1: Set the argument, x, on the keyboard. Print argument.

Carriage position 2: Set the sixth difference, D^{VI}, on the keyboard. Print D^{VI} and to D^{V} to form the new fifth difference.

Carriage position 3: Print the new fifth difference D^{V}. Add D^{V} to D^{IV} to form the new fourth difference.

Carriage position 4: Print the new fourth difference D^{IV}. Add D^{IV} to D^{III} to form the new third difference.

Carriage position 5: Print the new third difference D^{III}. Add D^{III} to D^{II} to form the new second difference.

Carriage position 6: Print the new second difference D^{II}. Add D^{II} to D^{I} to form the new first difference.

Carriage position 7: Print the new first difference D^{I}. Add D^{I} to f to form the new function value.

Carriage position 8: Print the new function value.

Carriage position 9: Carriage returns to position 1 and cycle repeats.

Allowances could be made to cope with instances when the differences changed sign but the basic method remained the same. Comrie (1936*b*) claimed that by using the National Accounting Machine in this way 200–300 values could be produced per hour. The use of this machine therefore again significantly eased the burden of table-making at the NAO.

Comrie's second important application of the National Accounting Machine to the work of the NAO was the converse of summation from finite differences: that is the differencing of functions up to the fifth difference. This application was extremely significant to the work of the NAO because it could be used as a powerful method of checking computed tables. Checking computed tables was a very important part of the NAO's work and was performed largely by differencing computed tables. Checking tables by recomputation was not only a costly procedure but also one which offered no guarantee against the repetition of any particular error. By differencing a function tabulated at uniform intervals of the independent variable until the differences oscillated in sign, the behaviour of a function could be examined. If an error had occurred in the tabulation of a function it propagated through the difference columns. An examination of those columns revealed not only whether or not an error had been made but also the position and magnitude of that error.

Because differencing required only trivial subtraction, it was of no advantage to use desk calculators and the work had previously been done by hand. Comrie found that by using the National machine it was possible to mechanically difference a function to fifths much more quickly than it could be done by hand. At each step in the cycle of operations successive difference columns were formed and stored in individual registers. At the end of each cycle the argument, the function, and the successive difference columns were printed. When the completed difference table had been produced the smoothness of the differences could be examined and any errors detected.

In addition to applying the National Accounting Machine to table-making and table-checking, Comrie also developed a method of interpolation based on Everett's formula using the machine and explored the possibility of using the machine for the formation of sums of products. Therefore at a cost of £500–£600 for the 6-register machine, the National was an important addition to the computing equipment of the NAO.[8]

Apart from introducing the National Accounting Machine into the NAO, Comrie continued to work on finding further scientific applications for punched card machines. Besides the preparation of tables of the position of the Moon computed using Brown's *Tables*, the only other calculation of any size which Comrie performed at the NAO using punched cards was the conversion of hundreds of polar co-ordinates to rectangular co-ordinates. This was a relatively simple process and involved using a Hollerith multiplier to form the necessary products. The work was carried out in 1933 on machines at BTM's premises in Victoria House, London; the task was not large enough to warrant the expense of reinstalling Hollerith machines at the NAO. Although Comrie did not use punched cards for any other major computation at the NAO, he was aware of the potential of the machines. He suggested, for example, that punched cards could be used to tabulate the apparent places of the stars which were published annually. Because this work was shared by four international offices the amount of computation carried out in any one country could not support a punched card installation in any one location.

By the mid-1930s, therefore, much of the computational work was performed with the aid of calculating machinery. Comrie's appointment to the Office had had a dramatic effect on the type of computational methods used there. However, the fact that the NAO used mechanized computing methods does not classify the NAO of the late 1920s and early 1930s as a computing centre in terms of the characteristics outlined in Chapter 1. Using desk machines did not mean that the NAO automatically began to take in work from other Admiralty establishments or elsewhere. To consider the NAO as computer centre it is necessary to look further than the type of computing methods it used for its own work and study the wider role, if any, that the NAO played in the late 1920s and early 1930s.

ENDNOTES FOR CHAPTER 3

1 Today this work is still shared between international offices although many organizational changes have occurred. In 1960 the British *Nautical almanac* and the *American ephemerides and nautical almanac* were unified into a single publication under the latter title. This title was changed in 1981 to *The astronomical almanac* which is again jointly published by both offices.

2 In addition to changing the computing methods used at the NAO, Comrie also redesigned the *Nautical almanac, abridged for the use of seamen* for 1929 and completely revised the *Nautical almanac* for 1931. A detailed discussion of Comrie's work in this area is outside the context of this book.

3 The crossfooter was an accumulating mechanism into which numbers from the keyboard could be added or subtracted.

4 The total key caused the value of the register and/or the crossfooter, whichever had been selected, to be printed and the register or crossfooter cleared. Sub-totalling caused the total to be printed but prevented clearing.

5 Although both Hollerith and Powers punched card machines were functionally very similar they operated on different principles. Comrie chose to use the Hollerith machines of BTM over those of Powers because they gave the user more control over the functioning of the machine. The Powers machines operated using prewired connection boxes supplied by the manufacturer to the user's specification. Hollerith machines could be wired up by the user for any task on site and hence offered greatly flexibility to scientific users. A contemporary description of the punched card machines available in Britain in the late 1920s and early 1930s is given by Comrie (1930*a*).

6 Other card capacities also existed. Eighty-column cards of the same dimensions became the most common card capacity but 34- or 60-column cards were also available.

7 In this assumption Comrie was not entirely correct. From 1960 the published positions of the moon, or lunar ephemerides, were calculated directly from Brown's theory using electronic computers without the use of the *Tables*. In 1954 corrections, based on new calculations, to the already published lunar ephemerides were released and subsequent tables were entirely calculated in this way. Thus Comrie's computations for the years 1960–2000 were superseded.

8 In the early 1930s a desk calculator, e.g. a Brunsviga-Dupla, could be purchased for about £70.

4

The Nautical Almanac Office as a computing centre and the founding of the Scientific Computing Service

THE NAO: A COMPUTING CENTRE?

The fundamental purpose of HM Nautical Almanac Office was to calculate and publish accurate ephemerides to aid navigation and to assist astronomers. These ephemerides were published annually as the *Nautical almanac*. Comrie's principle responsibilities as Superintendent of the NAO were the supervision and improvement of the publication of the *Nautical almanac*. As part of this work Comrie undertook a major revision of the *Nautical almanac*, incorporated in the 1931 *Nautical almanac* which had remained more or less unaltered since 1834. He rearranged, rewrote, and enlarged it to make it both more useful and easier to read. Comrie's mechanization of the computing methods used to prepare the *Nautical almanac* (described in Chapter 3) introduced other revolutionary changes for the NAO.

However, Comrie's interests were not restricted to the work of the NAO. He was also involved in astronomical societies, tablemaking, and mechanical computation generally. It was these interests, particularly the latter two, which turned the NAO under Comrie's superintendentship from a well-equipped but isolated institution into a renowned computing centre.

Compare the NAO's traditional role following Comrie's introduction of mechanized computing methods with the characteristics describing a computing centre given in Chapter 1. The NAO had, under Comrie's leadership, a well-trained computing staff operating desk machines. The computing staff made up approximately two-thirds of the entire NAO staff. The remainder were astronomers. The computing machinery which Comrie installed at the NAO was the most up to date available in the 1930s and Comrie continued to develop efficient numerical methods for use with them. Even so, these three aspects (experienced personnel, computing machinery, and the continuing development of numerical methods for use with the machines) are the only characteristics of a computing centre which the NAO of the 1930s fulfilled. The most obvious role of a computing centre is to serve the computational needs of a community of users. The NAO had no users; the computations it carried out were for its own use. It did not provide a computing advisory service; it undertook no computing machine research other than Comrie's work identifying which commercial machines could be applied to scientific work; and the NAO's library was not open, as a general rule, to anyone who wished to use it.

On the surface, therefore, the NAO of the late 1920s and early 1930s cannot be classified as a computing centre. However, if the work of Comrie himself is examined more closely a different picture emerges.

First, consider Comrie's involvement with astronomical societies. While still a student in New Zealand (1912–16) Comrie began to develop an interest in astronomy and consequently joined the British Astronomical Association (BAA). His links with the Association were strengthened when he formed the computing section of the BAA in 1920. Comrie was responsible for the first issue of the BAA *Handbook* and subsequently gave practical advice and guidance to later editors of this publication.

As a result of his position as Deputy, and later Superintendent of the NAO Comrie became an active member of the International Astronomical Union in 1928. In 1932 he was elected President of Commission 4 (Ephemerides) of the Union, a post which he held until 1938. Membership of the International Astronomical Union was very much connected with Comrie's work at the NAO. One of his major contributions was to identify that work on the apparent places of the stars was being duplicated in different parts of the world and he took the first steps towards ensuring that such work was carried out only once and that the results were made internationally available.

Comrie was also an active member of the Royal Astronomical Society. In 1921 he published the first of many articles in the *Monthly notices of the Royal Astronomical Society* (Comrie and Levin 1921). In February 1925, before he joined the NAO, Comrie contributed a report on logarithmic and trigonometrical tables to the 105th annual general meeting of the Society (Comrie 1925*b*), to be followed seven years later by a second report on mathematical tables generally (Comrie 1932*b*). These two reports reflect Comrie's wider interest in mathematical tables and computation outside the field of astronomy.

Mathematical tables, particularly their accuracy and ease of use, was a subject in which Comrie was deeply involved outside the NAO as well as within it. Greaves says of Comrie

> He always believed that the maker of tables not only should, but also could, provide tables that were substantially free of errors, and the tables bearing his name have set up a standard which can hardly be surpassed. But this was not all: he believed that it was his duty to give the user of tables every possible convenience, so not only were his tables checked and re-checked, but they were designed to be easy to interpolate and easy to read. . . . Comrie had an expert knowledge of typography and he applied

this knowledge, together with unrivalled experience and tireless attention to detail, to achieve something like perfection in presentation as well as perfection in accuracy. (Greaves 1953, pp.300–1)

Comrie's table-making work outside the NAO was carried out for, and with, a variety of people. For example, Comrie was Secretary of the British Association for the Advancement of Science Mathematical Tables Committee (BAASMTC) from 1929 to 1936. As such he had a considerable influence on the way in which the first six volumes of tables published by the Committee were presented. In addition, Comrie directly contributed to two of the early volumes of tables: he edited *Volume V: factor tables* (1935) and calculated *Volume VI: Bessel functions* (1937). Comrie also pressed the Committee members to use calculating machines in their table-making work. Comrie himself led the way by applying a National Accounting Machine to his own work for the Committee in 1931 before he had one installed at the NAO! Later the National Accounting Machine acquired by the Committee was housed at Liverpool University under J. C. P. Miller's supervision.

As well as undertaking work for the Mathematical Tables Committee Comrie was involved in other table-making projects. In 1930 he edited a new edition of Barlow's *Tables* (Barlow 1914) and in 1931 published a set of 4-figure logarithm tables with L. Milne-Thomson.

One of Comrie's most extensive table-making projects while Superintendent at the NAO was undertaken in collaboration with Johann Peters. Comrie and Peters met at the Astronomische Gesellschaft in Budapest in 1930 and discussed the possibility of producing a set of 12-figure tables of the natural values of the six trigonometrical functions to every second of arc and publishing nine volumes of 7-, 8-, and 10-figure tables. Peters supplied the pivotal values for the tables to Comrie who interpolated between them using the NAO Burroughs Class 11 machine in 1931. Because of publication difficulties only the 8-figure tables of the four principle trigonometrical functions (sine, cosine, tangent, and cotangent) were published (Comrie and Peters 1939).

Apart from table-making, the other great interest in Comrie's life was to 'spread the gospel of mechanical computation' (Massey 1952, p.100). Comrie was not content to develop machine computing methods for use at the NAO alone. He felt strongly that others should be informed of his discoveries. Comrie began publishing articles about the use of calculating machines as early as 1925 while he was still in the United States (Comrie 1925*a*) and continued to do so for the rest of his life. While at the NAO (1925–36) Comrie published at least 16 articles on the use of calculating machines for computation – only 5 of which directly described work undertaken at the NAO. Comrie did not restrict his work to scientific journals. Many of his publications were written for the business community. For example, Comrie contributed to the *Transactions of the Office Machinery Users Association* (Comrie 1929*b*, 1930*a*, 1932*a*) and in

Office management and the office manager (Comrie 1928*d* and 1928*e*). By contributing to encylopaedias Comrie also reached a wider audience still (Comrie 1929*c*, 1930*b*, and 1935).

Comrie also lectured frequently on the use of calculating machines. In 1929 he gave a series of six lectures at Imperial College, he also taught a 6-week course at the University of California in 1932, gave the Newmarch lectures at University College, London in 1933, and presented papers to the Royal Astronomical Society (Comrie 1932*c*, 1932*d*, 1932*e*, and 1932*f*) and the Royal Statistical Society (Comrie 1936*b*).

Because of the publicity Comrie's work received through his numerous publications, Comrie was consulted by many different people for advice on computing methods and the purchase of computing machines. In the mid-1930s Comrie was advising both the British Tabulating Machine Company (BTM) and Powers-Samas on the capabilities of their punched card machines. He also received requests from academics wanting to set up computing laboratories at their universities. In such cases Comrie was generous in offering his time to give advice and, in the case of Liverpool University at least, offered reprints of relevant articles to form the basis of a student library on mechanical computation (Grace 1934). Government establishments too acted upon his advice, for example the Ballistics Research Department at Woolwich installed a National Account Machining on his recommendation (Comrie 1932*a*, p.17).

In some cases Comrie gave practical assistance to those who came to him with a computational problem. For example at a lecture to the Royal Statistical Society at Cambridge in 1936 (Comrie 1936*b*) Comrie invited queries on the subject of mechanical computation. G. B. Hey, a mathematician working on an agricultural statistical investigation concerning optimum sample sizes of sugar beet, subsequently approached Comrie and explained the difficulties he was encountering. Comrie invited Hey, and his colleague H. G. Hudson, to the NAO at Greenwich to discuss the problem. Through Comrie's personal connections with BTM, for whom he undertook consultancy work, Hey and Hudson went to the BTM factory at Letchworth. As a result all their data was punched onto cards and they were able to use BTM E6/6 tabulators to form the required totals and sums of products by a summary multiplication technique. Because of the extent of the calculation, the tabulator which Hey and Hudson used was fitted with additional relays and distributors by BTM (Comrie *et al.* 1937).

Because there is little documentation concerning Comrie's private work it is difficult to pin-point whom Comrie assisted in other capacities. What is clear, however, is that during 1936 Comrie became involved with artillery survey problems (Comrie 1942). This work was undoubtedly stimulated by the worsening political situation in Europe in the mid-1930s and the approach of war. Comrie devoted a considerable amount of energy to them and devised powerful computing methods using the Brunsviga Twin 13z calculating machine. He used this machine for the determination of bearings and/or distances from a variety of positions, intersections, and resections. Comrie's work in this area was so successful that the War Office adopted twin machines and Comrie's methods in those artillery and survey divisions which regularly carried out such calculations. Comrie also produced a booklet (1938*b*) on the application of the Brunsviga 13z twin machine to the survey problems given in the 1932 HMSO *Manual of artillery survey*.

By the mid-1930s, therefore, Comrie was devoting much of his time to table-making, writing, and consultancy work outside of his duties as Superintendent of the NAO. These activities, coupled with the calculating machines Comrie had installed at Greenwich, brought the NAO very close to becoming an unofficial computing centre.

THE FORMATION OF THE SCIENTIFIC COMPUTING SERVICE

In August 1936 Comrie resigned from his position as Superintendent of the NAO. According to his obituarists (Greaves 1953, Massey 1952, Porter 1953), wife (P. B. Comrie 1984) and colleagues (Burrough 1983, Hey 1983, Miller 1979), Comrie was an impatient and forthright man who did not wait patiently while cautious civil servants debated particular points and yet he is also remembered as a kind and generous person. Relations between Comrie and the Admiralty were never smooth. Comrie felt that the Admiralty was deliberately putting obstacles in his way as he tried to modernize the computing methods used at the NAO. The amount of outside computing and consultancy work which Comrie performed was also a cause of dispute between him and the Admiralty. It was simply not done for a civil servant to undertake such work. Although none of his obituarists give reasons for his resignation other than poor relations between Comrie and the Admiralty, it seems likely that the increasing amount of outside work which Comrie took on contributed to bringing the disagreements to a head. That is, the use of the NAO as a computing service centre was not acceptable to the Admiralty.

Having resigned from the NAO, Comrie had still to make a living; he had a wife and young child to support. The amount of informal consultancy work Comrie had carried out while Superintendent at the NAO had proved that some sort of scientific computing centre was needed in Britain at that time. Comrie's widow, Phyllis, recalls that Comrie wanted to work for himself and that to set up an independent computing service was a natural step (P. B. Comrie 1984). In 1937, therefore, Comrie privately set up the Scientific Computing Service (SCS) and ran it, initially, from his home.

Through Comrie's connections with BTM (which retained his services as a consultant and something of a casual salesman at £500 per annum plus 10 per cent commission for work introduced into BTM's service bureaux[1]), the

SCS rented offices at BTM's service bureau at Clifton House, Euston Road, London. Although so closely connected to BTM, the SCS remained an independent organization and, more importantly, was not restricted to providing a computing service based on punched card machines alone. In fact the SCS was equipped with many different types of desk calculating machine.

The company accounts relating to the period in which the SCS was set up have not been preserved so it is impossible to confirm how Comrie raised the capital to back his new venture. The minutes of the British Tabulating Machine Company (1936–7, Minute 2924) record that BTM declined to give financial support to the SCS. Despite this Comrie found the money to establish the Scientific Computing Service as a limited company on 28 January 1938. Comrie and his wife were the major shareholders but a Mrs L. M. Stokes also held shares. Comrie invited G. B. Hey, his privately employed computing staff and others to join him and the service came into operation with a staff of 16. Soon afterwards the SCS moved into its own premises at 23 Bedford Street, London.

In 1938 Comrie formalized the role he saw the SCS playing in the scientific community. In a booklet describing the activities of the company he stated that the main purpose of the SCS was to provide a 'recognised organisation or institution to which the investigator could turn with assurance that he would find his language understood, and the confidence that the facilities of experienced computers and mechanical equipment could be put at his disposal' (SCS 1938*b*, p.3). There was, until that time, no one in Britain to whom scientists and statisticians working in government, industry, or the universities could bring computational problems.

The new company utilized a wide variety of machines. In addition to access to the full range of punched card machines available at BTM's premises in Clifton House, the SCS was equipped with several Brunsvigas (models 13z, 20, and Twin 13z), an electric Mercedes model 38 MSW, an electric Madas, an electric Facit, a 10-column electric Victor adding and listing machine with sexagesimal keyboard, a 10-column electric Continental adding and listing machine, a 12-column 6-register National Accounting Machine, and a 135-character typewriter for the preparation of printers copy. With such a wide range of equipment available and a trained staff the SCS offered an extensive list of services. These were

(1) scientific calculations generally, especially on a large scale, in cases where the worker concerned did not have access to or knowledge of the most efficient machines and techniques;

(2) table-making and interpolation on the National machine;

(3) mathematical statistics, e.g. correlation, variation analysis, and regression, using punched card machines for the formation of sums of products;

(4) analysis of market research questionnaires by tabulation using punched cards;
(5) Fourier analysis and synthesis: Comrie already had experience in this field with his 1929 work on Brown's *Tables*; there had also been suggestions as early as 1929 that the technique be used for X-ray crystallographic analyses (Hartree 1929);
(6) survey calculations not now restricted to military applications to include determination of longitude and latitude from star altitude observations, the adjustment of triangulations, the calculation of map projections and conversions between co-ordinate systems;
(7) the numerical solution of differential equations and evaluation of finite integrals in some cases utilizing the National to build up the integral from finite differences;
(8) to provide an advisory service on the type of machines to buy or use for a particular application—it was stressed that this advice was given impartially and that most machines suitable for scientific work were available for inspection at Bedford Square 'without the embarrassment of the presence of importunate salesmen' (SCS 1938*b*, p.5);
(9) provision of courses in mechanical computation arranged to suit individual requirements;
(10) to act as a publisher and specialist bookseller of mathematical tables and notes on calculating machines and methods;
(11) the preparation of printers copy and proof-reading;
(12) demonstrations and lectures on mechanical computation.

(Paraphrased from SCS 1938*b*.)

Thus by 1938 Comrie had defined his terms of reference and was providing a unique and valuable service. In its first two years the SCS performed most of their proposed activities. Some of the work undertaken during this time made use of Hollerith punched card machines and reflects the early association with BTM (for example, a market research analysis for Radio Luxemburg and an analysis of the Family Allowance Budget Survey). Comrie's earlier application of punched card machines to Fourier synthesis in the preparation of tables of the position of the moon led to the SCS performing an investigation into the use of punched cards for crystallographic work. Comrie's experience of computing navigational tables at the NAO also brought the SCS work in the preparation of Hugh's *Tables for sea and air navigation* (Comrie 1938*c*). Exhibitions too were not uncommon. For example, the SCS demonstrated at the British Association meeting at Cambridge in August 1938 and at the Royal Physical Society exhibition at South Kensington in January 1939. At such exhibitions the SCS usually displayed the flexibility of Hollerith machines by illustrating their use for the summation of products, for integration from sixth differences (using a rolling total tabulator), and the analysis of lines in spectra. The table-making capabilities of the National were also demonstrated at such exhibitions along with a variety of smaller, less specialized calculating machines.

In addition to these standard types of computation the SCS carried out a number of unusual and difficult scientific computations in the pre-war period. The most notable of these was the work the SCS carried out for Appleton and Weekes in the analysis of the height variations in the E-layer of the ionosphere (Appleton and Weekes 1939). A less well-known computation, but a no less significant one, was the solution of a series of second-order integro-differential equations in 1939 for Buckingham and Massey at University College, London which arose from their work on the scattering of neutrons (Buckingham and Massey 1941).

Comrie also offered a consultancy service. His letter headings stated that he was a consultant on 'Calculating Machines', 'Mechanical Calculation', and 'Mathematical Tables'. For example, in 1938 he was advising the Ordnance Committee of the War Office on the purchase of National Accounting Machines for the calculation of trajectories (Comrie 1938*d*) and assisting the Jute Research Institute of Calcutta on the range of commercial calculating machines available and their individual merits and costs (Comrie 1938*e*).

During 1937, 1938, and 1939 work for the SCS was plentiful and Comrie employed a staff of around 20 which, in between large jobs, was kept occupied doing large-scale triangulations for survey maps involving the solution of over 100 simultaneous equations on Mercedes calculators. On 3 September 1939 Britain declared war on Germany and over the following six years Comrie and the SCS were widely used by the scientific community and the service ministries. The SCS was the only commercially run scientific computing service to exist in Britain at the beginning of the Second World War.

The SCS can undoubtedly be called a computing centre. The SCS served anyone who had need of the skills it offered. Because it was run as a business venture Comrie did, of course, charge for the services supplied by the SCS. None the less there were occasions when Comrie made exceptions particularly if he himself were being asked to act in an advisory capacity (Comrie 1938*d*).

Comrie employed a mixture of both mathematicians and computing staff, generally young women, who were well trained in the use of desk machines—often by Comrie himself. His close links with desk machine manufacturers as well as with BTM and Powers-Samas ensured that Comrie was aware of new calculating machines as they came onto the market. The SCS carried out a wide variety of computations, as outlined above, some of which required numerical research. Comrie himself acted as a consultant on computing techniques, mathematical tables, and calculating machinery. The SCS also acted as a publisher, primarily of mathematical tables, and maintained an extensive library and bookshop.

Of the nine criteria used to describe computing centres outlined in Chapter

1, the only area of activity in which the SCS took no part was in computing machinery research. Comrie examined desk and accounting machines for their potential application to scientific work as they came onto the market, but was not involved in, and did not agree with, designing machines to serve specific computational problems.

L. J. Comrie was, therefore, the driving force behind the modernization of the Nautical Almanac Office and the foundation of the Scientific Computing Service. Under Comrie's leadership both institutions can be classed as computing centres. However, it is significant that the Admiralty does not appear to have approved of, nor perceived a need for, the NAO to operate as a computing centre. The very fact that the SCS was both operationally and financially successful in its first few years of operation indicates that its services were needed by the scientific community in Britain at that time.

ENDNOTE FOR CHAPTER 4

1 BTM ran punched card service bureaux all over Britain. These were punched card installations which carried out tabulations on a contract basis for anyone who wished to use them.

5

The influence of analogue machines in the 1930s: the Manchester and Cambridge differential analysers

In contrast to the growth in popularity of desk calculating machines during the late 1920s and the 1930s which Comrie helped to promote, the use of mathematical instruments by British scientists did not increase on anything like the same scale. Planimeters, integraphs, and harmonic analysers continued to be manufactured, but the mathematical laboratories set up within university departments during the late 1920s and early 1930s were, in general, equipped with desk calculators and not mathematical instruments.

A few scientists, however, found the use of desk calculators too slow or too tedious for the type of calculations they wished to undertake. For example, the solution of differential or integral equations, or Fourier analysis. There were two main reasons why some scientists held these opinions. Firstly they did not have access to or up-to-date knowledge of the full range of calculating machines available. Secondly they were unfamiliar with the numerical techniques needed to make effective use of desk calculating machines. Consequently several specialized analogue computing devices were built in Europe and the United States during the 1930s. However, few of the analogue computing machines built in Britain at that time enjoyed any measure of success.

BRITISH ANALOGUE COMPUTING MACHINES OF THE 1930s

The indices and abstracts for 1930–9[1] reveal that there were few British books or articles published during this period which described new or improved mathematical instruments. The literature clearly illustrates that interest in constructing mathematical instruments was greater in Europe and the United States than in Britain.[2] In the United States, for example, at the Massachusetts Institute of Technology (MIT) there was an established research program concentrating on the design of large mathematical instruments.[3] In Britain there were no such research programmes and very few large mathematical instruments were designed or built. Goldstine (1972) remarks that he found it curious that very few British physicists or applied mathematicians did any work on constructing calculating devices between Kelvin in the late 1800s and Hartree in the 1930s. Those machines which were built in Britain were designed by engineers and scientists who felt that the disproportionate amount of time they were spending on numerical work could be better applied to their own research.

The device built by E. C. Bullard[4] and P. B. Moon in 1931 to solve second-order differential equations serves as a good example of the type of instrument built in Britain during the early 1930s. Bullard and Moon were involved in geophysical research at Cambridge in 1931 and found the solution of second order differential equations which arose from their work 'extremely tedious' (Bullard and Moon 1932, p.546). Bullard and Moon were aware of Kelvin's work on a machine to solve second-order differential equations,[5] but considered that the mathematical manipulations needed to get the equations into to right form for use with such a machine involved an unacceptable amount of work. They therefore set out to build a machine to solve second-order differential equations which would require little manipulation of the original equations.

The Bullard and Moon machine consisted, essentially, of a conducting coil suspended in a magnetic field. The current supplied to the coil was varied with time by an operator following a curve traced on a drum. If the change in current was represented $f(x)=0$ then the movement of the coil represented a solution to the second order differential equation

$$\frac{d^2y}{dx^2} + f(x)y = 0 \qquad (1)$$

A mirror which reflected a spot of light from a fixed source was mounted on to the coil. The movement of the coil was recorded by a second operator tracing the movement of the light spot. The authors reported an accuracy of 'a few percent' (p.549).

Although the motivation for building the machine had been to reduce the amount of effort required to solve second order differential equations, it appears that a considerable amount of numerical work had still to be done not only to reduce equations to the appropriate form (as in equation 1) but also to manipulate the boundary conditions in order to minimize errors in the machine. The device also required two operators, one to follow the input function and one to follow the light spot. The principle source of error was caused by the understandable failure of the operators to accurately follow the input function or the light spot.

The Bullard and Moon machine displayed four characteristics typical of the kind of large analogue computing device invented in Britain during the late 1920s and early 1930s. First, the machine was designed and built by individuals rather than commercially manufactured by an instrument-maker. Secondly, the development of mathematical instruments or computing machines was not the main interest of the inventors; Bullard and Moon were physicists who suffered computing difficulties and made a direct attempt to solve them by the construction of a machine. Thirdly, as far as can be established, the machine was the only one of its kind to be build. Fourthly, the machine had no influence on future computing machine developments.

An example of a more widely publicized analogue calculating machine designed and built in Britain during the 1930s was the Mallock Machine. The machine was designed by R. R. M. Mallock in 1931 to solve linear simultaneous equations and used a series of interconnected transformers to represent the unknowns in a system of linear simultaneous equations. Coils on the transformers represented the coefficients in the equations: the number of turns in each coil could be adjusted to represent specific coefficients. Coils on separate transformers but relating to the same equation were interconnected to form a closed circuit. If the machine was set up to solve a set of three by three equations there would have been three such closed circuits. Alternating current was applied to the transformers and when the voltage levels of all the transformers had reached an equilibrium the solutions to the equations were read from the machine.

The Mallock Machine differed from the Bullard and Moon machine in several important respects over and above differences in application and construction. Like Bullard and Moon, Mallock had found the labour of numerical computation very laborious and designed a machine to overcome the problem. As in Bullard and Moon's case, the design of computing machines or mathematical instruments was not Mallock's main interest; Mallock was a demonstrator at the Cambridge University Engineering Laboratories. Unlike the Bullard and Moon machine which the inventors constructed themselves, the Mallock Machine was built, under Mallock's supervision, by the Cambridge Scientific Instrument Company in 1933. Although the agreement reached between Mallock and the Cambridge Scientific Instrument Company for the construction of the machine clearly shows that the company planned to manufacture further Mallock Machines, there proved to be no demand for the machine because of its inability to handle ill-conditioned equations.

Unlike any other British analogue machine of the early 1930s, the Mallock Machine was a well publicized device due, perhaps, to the financial interest the Cambridge Scientific Instrument Company had in the machine. The Mallock Machine was exhibited at the Royal Society Conversazione and the International Congress of Applied Mathematics in 1933 and 1934 respectively and four descriptions of the device appeared in the contemporary literature.[6] Because of the publicity the Mallock Machine received it was a fairly well known machine and has very often been included in reviews of analogue computing machines which have since been published.[7]

The machine was also used experimentally by several people. The most notable of these were staff of the Royal Aircraft Establishment at Farnborough and A. C. Aitken, the Edinburgh mathematician. The use of the Mallock Machine by scientists from outside Cambridge contrasts sharply with other analogue machines of the period, such as the Bullard and Moon integrator, which were not so well known.

Again unlike any other British analogue calculating device of the early 1930s, the Mallock Machine directly influenced the construction of another device. In 1936 a patent was taken out by Electrical Improvements Ltd and C. Blackburn for a network calculator based on Mallock's construction of a series of interconnected transformers. Although the Blackburn Network Calculator was built to simulate electrical power networks, it could be applied to more general problems involving the solution of simultaneous equations. Like most of the other analogue instruments built in Britain in the 1930s, the Blackburn Network Calculator did not have an influence on scientific computation generally.

THE MANCHESTER DIFFERENTIAL ANALYSER

The only class of analogue machines to be used successfully in Britain during the 1930s were the differential analysers.

In 1930 V. Bush built a machine at the Massachusetts Institute of Technology called a differential analyser. The machine was designed to solve differential equations up to the sixth order and consisted of six integrating units connected together by bus shafts (Bush 1931). Each integrating unit was made up of a horizontal disk with a vertical wheel resting upon it. The output shaft from each integrating unit was connected to the bus system through a torque amplifier. The torque amplifier allowed each integrator to exert sufficient force onto the bus system to drive other components in the machine without causing slip between the integrating disk and wheel. Input tables, output tables, and multipliers were also attached to the bus system.

In constructing the machine one of Bush's aims had been to produce a flexible and rugged machine which operated to a reasonable precision. Bush gave a precision of one part in one thousand in each of the integrators and stated that the expected overall precision would be slightly lower than this. The machine was successful in achieving these aims and several copies based on Bush's original design were constructed in other countries. The first differential analyser to be built in Britain was a model of Bush's machine constructed in 1934 by D. R. Hartree.

Hartree (1897–1958) had already had a considerable experience of numerical computation. During the First World War Hartree had been heavily involved in A. V. Hill's work on anti-aircraft ballistics. Much of the work had involved the solution of large numbers of differential equations using desk calculators. Hartree became an expert in this type of numerical work and introduced new techniques into the field. After the war Hartree returned to Cambridge to finish his undergraduate studies. He remained at Cambridge until 1929 by which time he was a fellow of Christ's College. Hartree's main interest was in the distribution of electron density in atoms; a field in which the knowledge of computational techniques he had gained during the war was invaluable.

In 1929 Hartree became Professor of Applied Mathematics at Manchester University where he applied wave mechanics to the study of electron densities in atoms. This involved the determination of wave functions by the numerical solution of differential equations. Although in most cases the calculations required were not difficult, they were extremely tedious to carry out for large numbers of high-order differential equations. Darwin, Hartree's biographer, states that the Bush differential analyser greatly interested Hartree who immediately saw how it could alleviate the amount of numerical computation he regularly performed.

In 1933 Hartree visited Bush at MIT and took the opportunity to use the differential analyser for his work on the approximation of the atomic field and wave functions of mercury. On his return to England, Hartree continued the work manually using desk calculators and found that the immense amount of labour required to complete the calculations emphasized the value of the differential analyser. In order to demonstrate the principles of the differential analyser Hartree built a Meccano[8] model of the Bush machine. Hartree recalled that

> the first results were successful beyond my expectations and suggested that it would be practicable to build such a model to do serious work on problems for which high accuracy was not required in the results. (Hartreee 1940*a*, p.160)

As a result of his successful experiments, Hartree and a research student, A. Porter, build a complete four-integrator model differential analyser largely from standard Meccano parts. The model differential analyser cost approximately £20 to construct (considerably less than the cost of a Brunsviga of the same period); the machine is fully described by Hartree and Porter (1935).

The model differential analyser was a useful machine and original work was performed using it. Soon after it was completed Porter used the machine to determine the approximate wave functions for chromium as part of his M.Sc. thesis and hence demonstrated the usefulness of the differential analyser. Hartree then secured a gift from Sir R. McDougall, the deputy treasurer of the University, to fund the installation of a full-sized differential analyser at Manchester. With the help and advice of Bush, who supplied drawings of his original machine and suggested possible improvements, Hartree arranged for the construction of a four-integrator differential analyser by the Metropolitan-Vickers Electrical Company, Ltd. In the period prior to the delivery of the differential analyser Hartree prepared for its arrival by using the model machine as an experimental device with which to gain experience in the preparation and setting up of equations for solution on a differential analyser.

The full-sized four-integrator differential analyser was installed in the basement of the Physics Department of Manchester University and was formally opened in March 1935 (Plate 10). Almost immediately McDougall presented a second gift to the university for the construction of a further four integrators for the differential analyser, which brought the total donated to £6000. The completed machine had eight integrating units and could therefore be applied to ordinary differential equations up to the eighth order.

The differential analyser was very much Hartree's machine which he ran with the assistance of Porter, interested members of staff, and research students. Despite the absence of any formal operating staff, the Manchester differential analyser was well used by many members of the university.

Hartree had primarily installed the differential analyser for work on electron density distribution in atomic fields but the machine was never used for this purpose because in the period just before the machine was installed, discoveries by Hartree and Fock on the behaviour of electrons led to a reappraisal of the method used to find electron distributions. The new technique did not require the use of the differential analyser. None the less, Hartree applied for and was awarded a Department of Scientific and Industrial Research (DSIR) grant for the further development of the differential analyser at Manchester and the investigation of problems applicable to it. Hartree, with his research assistants and students, used this grant to carry out a wide range of calculations including several industrial problems. In cases where industrial research work was carried out on the differential analyser Hartree liaised with staff from the companies concerned. One particular investigation concerning potential distribution in cylindrical thermionic valves involved three different companies, and was undertaken by Hartree in conjunction with staff from the M.-O. Valve Co. on behalf of the General Electric Company Ltd. and Marconiphone Ltd.[9]

Additional units for the differential analyser were made specifically for particular types of calculation. For example, for one problem concerning computations investigating time-lag in control systems carried out for ICI (Alkali) Ltd, a special purpose input table was constructed and attached to the machine.[10] Some of Hartree's colleagues at Manchester also took an interest in the differential analyser and began to use it in their work or to develop attachments for it. P. M. S. Blackett, professor of physics, became interested in the technical side of the differential analyser and, with F. C. Williams[11] of the Electro-technics Department, developed an automatic curve follower for the machine.

Hartree also carried out investigations chosen specifically because they required an unusual application of the differential analyser. One such computation was an analysis of train running times. This application required discontinuous data which represented track gradients and speed limits, therefore counters, rather than the standard graphical tables, were used as input devices.

Before the war the machine was therefore used in a variety of applications by many people. Darwin summed up the position of Manchester differential analyser. He wrote

> With his invariable generosity Hartree gave service with it to a very great number of enquiries, and the service included giving them his intimate experience of the general way that numerical problems could be solved. (Darwin 1958*b*, p.107)

Consequently, although the Manchester differential analyser was not initially installed in order to provide a computing service, the DSIR grant, which provided Hartree with the financial support needed to maintain the machine, meant that in practice it was used as an additional computing facility for those British research scientists who approached Hartree from industry, government, or universities. At this time, 1935–9, there was no other full-size differential analyser in Britain and a specialized computing centre developed around the Manchester machine. The group had little formal organization but scientists who needed faster methods of solving complex or high-order differential equations came to Manchester to use the machine.

Hartree also published descriptions of the differential analyser and explained what type of problems the machine was capable of solving. Between 1935 and 1940 he published four accounts of differential analysers[12] and co-authored several papers describing application which arose from the research work carried out on the machine.[13] This publicity, and the respect Hartree commanded as a mathematician, encouraged scientists to come to Manchester to use the machine. The success of the Manchester differential analyser brought an increased awareness in the need for large-scale computing devices to be made available to scientists and led to the construction of several model differential analysers at several institutions. The earliest of the model machines was built in Cambridge and resulted in the installation of a second full-size differential analyser in Britain.

THE CAMBRIDGE DIFFERENTIAL ANALYSER AND THE FOUNDATION OF THE UNIVERSITY MATHEMATICAL LABORATORY

In 1932 John Lennard-Jones took up the Plummer Chair of Theoretical Chemistry at Cambridge University and began to establish a school of theoretical chemistry there. Lennard-Jones had previously worked on the interatomic forces of gases and solids, first as a postgrauate student under R. H. Fowler, then at Bristol University as Reader in Mathematical Physics and later Professor of Theoretical Physics. The main interests of Lennard-Jones's theoretical chemistry group at Cambridge concerned the behaviour and distribution of interatomic particles. By 1935 the work of the group had reached a point where the amount of computation necessary, primarily the solution of differential equations, was becoming too large and too complex to be carried out by hand. According to Mott (1955), his colleague and obituarist, by 1935 it had become obvious to Lennard-Jones that some form of mechanical computing aid was necessary. Hartree and Porter's work on the differential analyser at Manchester was well-reported in the contemporary literature and Lennard-Jones saw that a differential analyser could be applied to the differential equations arising from his own work.

Lennard-Jones approached Hartree about the possibility of building a model differential analyser at Cambridge. Hartree and Porter offered help and advice to Lennard-Jones based on their own experience in constructing the Meccano differential analyser at Manchester and suggested ways in which a model machine might be improved. As a result a Meccano differential analyser was built in 1935 in the Physical Chemistry Department of Cambridge University by J. B. Bratt under the supervision of Lennard-Jones.

Although Lennard-Jones's model differential analyser was based heavily on Hartree's model machine at Manchester it was a more accurate and reliable machine. The most important difference between the Manchester and Cambridge model differential analysers was the design of the torque amplifiers. The torque amplifiers gave the output shaft of each integrator enough power to drive the other mechanisms within the machine. The torque amplifier which Hartree had built was a simplified version of the one stage device used by Bush; it had an amplification factor of approximately 80. Hartree had also used ground glass integrating disks to increase the friction between the disk and the integrating wheel and hence provide a greater initial torque. When the fourth integrator was added to the Manchester model, the torque amplifier was redesigned as a two-stage device to improve amplification to an order of approximately 2000. Lennard-Jones adopted the two-stage torque amplifier design but further improved the amplification. All the integrating disks on the Cambridge model were made of plate glass as the added friction afforded by ground glass was no longer necessary. Apart from the torque amplifiers there were several other, less important, differences which were mainly concerned with changes to the arrangement of components.

The model machine was situated in the Physical Chemistry Department at Cambridge University for use by the theoretical chemistry group, but other members of the university soon began to take an interest in the device. In particular the machine attracted the attention of M. V. Wilkes, a research student in the Cavendish Laboratory at Cambridge.[14] Wilkes attended a demonstration of the model in March 1936 and saw that a differential analyser could be used to solve the differential equations which resulted from his work on the propagation of radio waves. Wilkes expressed his interest in using the differential analyser for his work and Lennard-Jones gave Wilkes permission to use the machine. When Bratt left Cambridge in late 1936 Lennard-Jones offered Wilkes the job of providing technical assistance to other users of the differential analyser. The job paid a small salary and gave Wilkes access to the machine which he used to finish his Ph.D thesis. Unlike Hartree, who was very closely associated with the construction and operation of the model differential analyser at Manchester, Lennard-Jones was not personally involved with the running or maintenance of the Cambridge machine. Having initiated the project and obtained the necessary funding Lennard-Jones left the day-to-day running of the model differential analyser to Bratt and, later, to Wilkes.

On seeing the success and usefulness of the model machine, Lennard-Jones began to press the University authorities to install a full-size differential analyser at Cambridge. He was not, however, working for the installation of a large differential analyser for the use of the theoretical chemistry group alone. Lennard-Jones recognized that such a facility should be made *officially* available to the research staff of Cambridge University as a whole. This formal organization of resources differs from the situation at Manchester where the differential analyser was used by scientists from inside or outside the university who directly approached Hartree on a personal basis.

Lennard-Jones was, moreover, concerned that something more than an isolated differential analyser was needed by Cambridge scientists. He wanted to establish a computing laboratory to house the model differential analyser, a full-size differential analyser, and a range of other computing machines. The laboratory was to be equipped with a sufficiently large range of calculating machines to perform the scientific computations generated by all university departments. Lennard-Jones was thus proposing to centralize scientific computation within Cambridge University.

In December 1936 Lennard-Jones succeeded in getting the University Faculty Board of Mathematics to prepare a report on the need for a computing laboratory at Cambridge (University of Cambridge, Faculty Board of Mathematics 1936). This report presented a general case for the establishment of a computing laboratory. Despite Lennard-Jones's involvement there is nothing in the preamble describing the university's need for computation to suggest that the driving force behind the document came from the theoretical chemistry department. The report first discussed the need to alleviate the long laborious calculations carried out by astronomers and went on to present the case that all physical scientists were encountering problems which could not be solve analytically and thus required arduous numerical work.

The report noted the recent improvements to desk calculating machines and their increased use for scientific work. At the same time it was remarked that the number of desk machines in Cambridge was very limited and this situation could be improved at a relatively low cost. The recent application of punched card machines to harmonic analysis[15] and the development of linear simultaneous equation solvers (particularly the Mallock Machine at the Cambridge Engineering Laboratories) were also cited to demonstrate the advances in mechanical computation over the preceding decade. The Bush differential analyser, Hartree's work at Manchester, and the existing model at Cambridge were then discussed and the case for a full-size differential analyser at Cambridge presented.

The Faculty Board of Mathematics stated that it was satisfied that the model differential analyser had proved the need for the installation of a full-size machine at Cambridge. It dismissed the suggestion that it would be better to employ a staff of trained computing staff equipped with desk calculators on the grounds that the differential analyser could give reliable results to complex, laborious problems in a few days while it would take many weeks or months to do the same work using desk machines. If desk machines were the only calculating devices available many exploratory calculations would not be carried out as the time scale involved would have been too long to be useful. Hence the differential analyser could be used as a convenient research tool opening up problems which would otherwise remain unexplored. The Faculty Board of Mathematics concluded that the installation of a full-size differential analyser would be more useful than a large computing staff in assisting research at Cambridge and considered it 'essential in the interests of the physical sciences in Cambridge to develop a computing laboratory' (University of Cambridge, Faculty Board of Mathematics 1936, p.4).

The report of the Faculty Board of Mathematics went on to make definite proposals concerning the organization and cost of such a laboratory. It was suggested that the laboratory be attached to the Mathematics Faculty and set up close to both the Mathematical Institute and the Cavendish Laboratory. The laboratory was to be equipped with a full size 8-integrator differential analyser to be extended as soon as possible to a 12-integrator machine, the model differential analyser, a Mallock simultaneous equation solver, and a wide range of desk calculating machines. It was estimated that the initial cost of equipping the laboratory would be £9000. A technician to maintain the machines was to be employed also. Research workers using the laboratory would operate the machines themselves under the supervision of the technician and the laboratory director who, it was recommended, should be a professor from the Mathematics Faculty. It was planned that a reserve fund for the laboratory be created to finance any machine developments which took place as research scientists used and gained familiarity with the machines. The Faculty Board of Mathematics was prepared to donate £2500 to the creation of the laboratory.

The report of the Faculty Board of Mathematics was then circulated to the other faculties in the School of Physical Sciences (Physics, Chemistry, Geology, Geography, and Engineering). These faculties approved the scheme to set up a computing laboratory and the Council of the School of Physical Sciences prepared a report which it presented to the General Board of the University (University of Cambridge, General Board 1937*a*). This report agreed with, and was essentially the same as, the report of the faculty Board of Mathematics with a few exceptions and additions. For example, the Council called for the installation of a National Accounting Machine and a significant increase in the number of desk machines for the laboratory. It was also suggested that a lecturer attached to the Mathematics Faculty should be assigned, part time, to the laboratory to take responsibility for the desk machines and lecture on numerical computation, thus extending the role of the laboratory to include undergraduate teaching.

The Council also made further suggestions as to how the money needed to set up the laboratory was to be found. It was proposed that most of the money was to come from the development funds of faculties using the laboratory. This was to be supplemented by the £2500 promised from the Mathematics Faculty and £1750 donated by the Engineering Faculty to purchase the Mallock Machine which was then situated in the Engineering Laboratories but which technically belonged to the Cambridge Scientific Instrument Company, Ltd. Besides the School of Physical Sciences, the Departments of Economics and Politics, Biology, and Agriculture also expressed an interest in using a computing laboratory and gave their support to the project.

On 23 February 1937 the University of Cambridge approved, in principle, the creation of a computing laboratory and over the next few months moves were made to achieve this. After much discussion it was decided to name the computing laboratory the 'Mathematical Laboratory'. This title was chosen as 'non-committal' and allowed for the use of both desk calculators and the differential analyser in the laboratory. Lennard-Jones, along with others, felt that the words 'computing' or 'calculating' did not adequately describe the analogue nature of the differential analyser. It was also anticipated that the name of the laboratory should be general enough to allow computing machine developments to take place. The final plans for the Mathematical Laboratory were approved in May 1937 and the Laboratory opened the following October.

Lennard-Jones was appointed part-time director of the Cambridge University Mathematical Laboratory and M. V. Wilkes took up a post in the laboratory as university demonstrator. Wilkes was responsible for supervising the construction of the full-size differential analyser and visited Hartree in Manchester to learn what he could about the machine there. Metropolitan-Vickers undertook the construction of the full-size differential analyser but progress was slow due to the increasing number of military contracts which the company was obliged to take on. The machine was not delivered to Cambridge until late 1939 after the start of the war.

Initially the Mathematical Laboratory did not have any accommodation. As a result, the machines attached to the laboratory remained where they had been previously. The model differential analyser was situated in the theoretical chemistry department; the Mallock Machine (bought by the University in June 1937) was located in the engineering laboratories; and the few desk calculating machines which had been allocated to the laboratory were scattered throughout the university. In addition a National Accounting Machine had not been purchased. Although the Laboratory had been given accommodation in the Old Anatomy School, a significant amount of work needed to be done to refurbish the building before the Mathematical Laboratory could move in. Thus, although active from October 1937, the Mathematical Laboratory did not physically come together in one building until October 1939. By this time war had been declared, Wilkes had left Cambridge for war service and the Ministry of Supply had requisitioned the Laboratory's facilities for the duration of the war.

In principle, therefore, the Cambridge University Mathematical Laboratory was the first mathematical or computing laboratory to be established as a computing centre for all the scientists working in a particular university. Because of accommodation problems, the delay in the arrival of the differential analyser and the outbreak of war it did not achieve its aims in practice until after the war. Nevertheless, the reports of the General Board and the Faculty Board of Mathematics clearly show that the concept of centralizing computing power within a university was well understood in Cambridge by 1936. The laboratory was intended to serve the University as a whole; it was to be equipped with an extensive range of computing devices; it was anticipated that research would be undertaken to increase the computing power of the machines in the laboratory. The fact that Wilkes and a technician were the only full-time staff attached to the laboratory indicates that the field of numerical analysis was not yet perceived as a separate discipline to the experimental and theoretical sciences.

OTHER DIFFERENTIAL ANALYSERS IN BRITAIN

There is no doubt that Hartree's work at Manchester led to the construction of the model differential analyser at Cambridge and, subsequently, to the establishment of the Cambridge University Mathematical Laboratory in 1937. However, during the late 1930s and early 1940s several other model differential analysers were built in Britain. The most well-known of these was the model built in the Physics Department of Queen's University, Belfast by H. S. W. Massey and his colleagues. Massey, who was aware of both Bush's work in the United States and Hartree's work at Manchester, felt that a differential analyser was a valuable computing tool. Because of the great cost of installing a full-size machine Massey decided to construct a model differential analyser and contacted Hartree and Porter in Manchester. With help and advice from Porter a four-integrator model differential analyser was built. It was not not constructed from Meccano but from parts obtained or made by the Queen's University Science Workshops. The machine cost approximately £50 and was, Crank (1947) suggests, a much less accurate device than the model differential analyser built at Cambridge.

In 1938 Massey was appointed Goldsmid Professor of Mathematics at University College, London and brought the differential analyser with him. There was no suggestion at Queens in Belfast or at University College, London that the model was a demonstration device intended to illustrate the need for a larger machine. The model differential analyser was for the use of Massey and his colleagues. During the war Massey's model differential analyser was destroyed during an air raid.

In several cases work done by Hartree on behalf of a government research establishment or company had led to the construction or acquisition of other model differential analysers by the institution concerned. For example, staff at the Coast Artillery Experimental Station and at the General Electric Company Ltd constructed model differential analysers after working with Hartree at Manchester. At the beginning of the war J. R. Womersley, later the first Superintendent of the National Physical Laboratory Mathematical Laboratory, and who had previously worked with Hartree on the mechanical solution of partial differential equations, began to build a machine for the Armanent Research Department at Woolwich. It, like Massey's machine, was destroyed. In the early 1940s the Valve Research Department of Standard Telephones and Cables Ltd installed a model differential analyser which had originally been built by an actuary, R. E. Beard, to solve differential equations which arose from his work in insurance. Another three model differential analysers are known to have existed. One was built by a Macclesfield schoolmaster, a second was constructed in the Physics Department of Birmingham University, and a third was developed before the war in the War Office Projectile Development Establishment, Fort Halstead.

In the 1930s, therefore, the only large analogue computing devices to be in any way influential were the differential analysers at Manchester and Cambridge Universities. At Manchester the model and full-size differential analysers were well used in an informal way. Only at Cambridge did the differential analysers prompt the idea that a computing centre would be a useful common tool for Cambridge research staff. The outbreak of the Second World War in 1939 postponed the development of the Cambridge Mathematical Laboratory as computing resources and university personnel became involved in the war effort.

ENDNOTES TO CHAPTER 5

1 *Subject Index to Periodicals*, The Library Association, London: The *British Museum Subject Index*; *Science Abstracts – A Physics*, Institution of Electrical Engineers.

2 This literature search revealed published descriptions of only 4 analogue

computing instruments designed in Britain during the 1930–9 period whereas at least 30 American and European machines were discussed.

3 From the mid-1920s onwards a group working under Vannevar Bush in the Electrical Engineering Department of MIT designed several different large analogue computing instruments including the network analyser, the product integraph, the differential analyser, the photoelectric integraph, and the cinema integraph.

4 Bullard later became director of the National Physical Laboratory (NPL) 1950–5.

5 Thompson and Tait's *Treatise on Natural Philosophy* 1890 was a very well-known text; pp.947–9 of Part I relate specifically to this type of device.

6 Mallock 1933; Anon. 1933; Anon. 1934; and Anon. 1935*d*.

7 See for example Lilley 1942; Comrie 1944; Frame 1945; Murray 1948; Hartree 1950; Engineering Research Associates 1950; and Goldstine 1972.

8 Meccano was a constructional toy widely sold in Britain.

9 Crank *et al.* 1939.

10 Callender *et al.* 1936; Hartree *et al.* 1937.

11 Williams was later responsible for computer development at Manchester University.

12 Hartree 1935; Hartree 1938; Hartree 1940b.

13 For example, Hartree and Ingham 1938; Hartree and Porter 1938; Hartree, Porter, Callender, and Stevenson 1937; Callender *et al.* 1936; Copple *et al.* 1939; Crank *et al.* 1939; Hartree *et al.* 1936.

14 In the late 1940s Wilkes became a leading pioneering in the computing field by developing the EDSAC computer at Cambridge University.

15 For example by Comrie at the National Almanac Office.

6

The Second World War: The emergence of government computing centres

The work of Comrie, Hartree, and Lennard-Jones during the 1930s had done much to develop both calculating machines and computing techniques. Hartree and Lennard-Jones had also been instrumental in providing computing resources for particular groups of scientists. However, by 1939 the only organization to which scientific workers in government research establishments could apply for computational advice or practical assistance was Comrie's Scientific Computing Service (the SCS).

COMPUTATION WITHIN THE SERVICE MINISTRIES, 1939–1942

The advent of the Second World War brought an increased demand for computation in terms of both bulk and mathematical complexity. This increase was most pronounced in two particular areas, first, trajectory calculations and secondly applied research into new forms of weapons and defence. In the first case, in the period of rearmament directly preceding the war and, more urgently, after the outbreak of war in 1939, an effort was made to provide the armed services with ballistic tables and trajectory data for all types of weapons then in use. To try to deal with the heavy load of computation this required, the War Office and the Air Ministry enlisted the help of both the SCS and the NAO to compute the tables. But this was not sufficient: further research was needed to find more rapid methods of calculating trajectories.

The second expansion in computing requirements experienced by the service ministries arose as a result of the acceleration in applied research concerning weapons, ships, aircraft, and defence systems prompted by the war. This affected most government research establishments and, in the majority of cases, was a problem which remained unsolved until after the war. Within certain organizations, however, efforts were made to find alternative ways of performing the necessary calculations.

The only readily available and well-equipped computing centre to which scientists could apply for computational assistance was the SCS. Details about the contribution the SCS made to the war effort are difficult to verify. There are few surviving accounts of the work the SCS actually performed partly due to the classified nature of much of the work carried out during the war and partly due to the scarcity of records Comrie left behind.[1]

Before the war Comrie had performed triangulation calculations for the War Office and throughout the war the War Office continued to use the SCS. Comrie was fond of boasting that within three hours of war being declared War Office officials approached him with an urgent request for the computation of a set of ballistic tables for the control of certain types of anti-aircraft gun in which predictors[2] were not used. War Office workers had asked for three months to complete the three sets of tables required. In 24 hours Comrie had planned and begun the first of these tables and undertaken to prepare a second. In 12 days the two sets of double entry tables had been computed on the SCS National Accounting Machine, proof-read, printed, and bound. War Office staff computed the third set in three weeks (Comrie 1948*b*).

The SCS also carried out triangulation work for the conversion of Prussian latitude and longitude co-ordinates to rectangular co-ordinates used by the British armed forces, and an investigation into Kine Theodolite calculations[3] for the War Office. SCS war work for other government bodies included computations for Air Intelligence in 1940 to establish the location of German radio guidance transmitters, a calculation concerning the degaussing of ships, the preparation of wind graphs for sound ranging analysis, ballistic tables for the United States Air Force, and the determination of the crystalline structure of penicillin for the Medical Research Council.

Most of the SCS's work was carried out at its premises in London. However some of the larger jobs which required the use of punched card machines were performed on whatever machines were available at the time. For example, SCS work was carried out using the Customs and Excise Hollerith installation at Buxton and machines belonging to the Milk Marketing Board at Cirencester. The SCS also used punches situated at the Ministry of Aircraft Production's administrative installation in Mill Bank.

Although the SCS was widely used by government departments during the war, it was not funded by the government and had to operate on a commercial basis. The terse minutes of the company board meetings do not contain many financial details (the early accounts of the company are poorly preserved), but certain conclusions can be drawn from the contents of the minutes. Before the war the company was financially stable and reasonably profitable. The increasing amount of war work the company was asked to undertake meant that the company rapidly expanded in terms of both staff employed and machines acquired. At the same time invoices for work performed were not being cleared quickly. The result of these two factors was a cash flow problem for the company. To try to relieve the problem Comrie is recorded as having waived his salary rights on several occasions. By 1944 these problems had been largely overcome, payments had been made and the company was again in profit.

At the beginning of the war the only service ministry to make special provision for the increased demand for scientific computation was the Ministry of Supply. The Ministry of Supply was created in August 1939 and took over the supply departments of the War Office and the Air Ministry. It also incorporated the Ordnance Board and the Research and Design Departments of the War Office. The Ministry of Supply was therefore responsible for armament research as carried out in the Research Department of the War Office and the ballistics research carried out by the External Ballistic Department of the Ordnance Board. In 1938 revisions were made to the theory concerning the air resistance of cylindrical projectiles which needed to be urgently incorporated into the trajectory calculations and ballistic tables of both new and old weapons if effective use was to be made of the available weaponry. Having to recompute ballistic data for weapons already in use disproportionately increased the amount of computation required. Although the Ordnance Board had installed several National Accounting Machines in 1938 specifically for calculating trajectories, these were not sufficient to meet all computational requirements.

In August 1939, as the political situation in Europe worsened, the Directorate of Scientific Research of the Ministry of Supply arranged for Lennard-Jones to form a research team attached to the External Ballistics Department of the Ordnance Board to be based at the recently created Cambridge University Mathematical Laboratory. Lennard-Jones negotiated with the university authorities on behalf of the Ministry of Supply to take over the Cambridge Mathematical Laboratory under a lease agreement. On 1 September 1939, two days before war was declared, Lennard-Jones began work at Cambridge for the External Ballistics Department (later part of the Armaments Research Department) of the Ministry of Supply. Under Lennard-Jones's direction a team was built up to carry out ballistics calculations for guns and rockets, research into high explosives theory, and sound ranging. The Mathematical Laboratory was equipped with the model differential analyser already housed in the Laboratory, the full size differential analyser (delivered to the Laboratory by Metropolitan-Vickers in late 1939), the Mallock Machine, and a range of desk calculators.

Although the Ministry of Supply External Ballistics Department needed to expand its computational capacity generally, the principal attraction of the Cambridge Mathematical Laboratory was the full-size differential analyser. The concept of a shared computing resource on which the Mathematical Laboratory was based was of little or no interest at this critical time. However, it was soon found that the accuracy of the differential analyser was too low for the routine calculation of trajectories.[4] None the less, the machine was used successfully for unusual experimental ballistics work. By 1942 the laboratory was generating sufficient mathematical problems to keep the differential analyser in operation 12 hours a day. Despite the number and scale of computations being performed on the differential analyser only a small staff, between two and four, worked full time on the machine. The bulk of the External Ballistics Department staff based in the laboratory used desk machines for their work.

The Cambridge Mathematical Laboratory primarily performed ballistics based work and seldom took in computations from outside the Ministry of Supply. Access to the Cambridge facilities by outside workers was occasionally permitted: for example, de Havillands' aircraft manufacturers used the Mallock Machine during the war. During the war the Cambridge Mathematical Laboratory was predominantly used by a single research team; the status of the laboratory as a computing centre was, therefore, lost.

Hartree's differential analyser group at Manchester also became attached to the Ministry of Supply at the beginning of the war. Initially the connection was an informal one in which Hartree took on a consultative role but in 1942 the group became part of the Headquarters Section of the Ministry of Supply and became known as SR(A). The Manchester differential analyser group was very different to that at Cambridge. At Cambridge the differential analyser was part of the computing facilities used by the External Ballistics Research group based there. At Manchester the differential analyser group acted primarily as consultants to a wide range of people inside and outside the Ministry of Supply. At first the Manchester group performed primarily ballistics calculations, including a number for the External Ballistics Department of the Ordnance Board before the Cambridge differential analyser became fully operational. Later the Manchester differential analyser was used for the calculation of heat flow in rocket tubes, for studying servomechanisms particularly in relation to fire control equipment,[5] the propagation of blast waves and many other military applications (Hartree 1947*a*).

The group working at Manchester was quite small. At the beginning of the war it consisted of Hartree and two others. Hartree, however, left the group to take up a post at the Projectile Development Establishment at Fort Halstead, which also came under the control of the Ministry of Supply. Although the staff at Manchester was reduced to one in 1940, by 1942 it had risen to four. In 1941 Hartree was transferred to Ministry of Supply Headquarters and was again able to supervise, and sometimes contribute to, the work performed on the Manchester differential analyser. Also in 1941 Hartree formed a second group at Manchester which used the differential analyser for its work on the development of the radar magnetron (Hartree 1947*a*). This group was attached to the Admiralty but reached an agreement with the Ministry of Supply which allowed it access to the differential analyser.

Although the Cambridge Mathematical Laboratory and the Manchester SR(A) group were affiliated to separate Ministry of Supply departments, it is important to note that both groups liaised with each other and with other groups working in the same field. For example, the Cambridge and Manchester groups both worked towards finding more rapid methods of calculating trajectories using desk machines in co-operation with other branches of the External Ballistics Department and the collaboration led to a significant improvement in the use of desk machines for ballistics calculations. The Cambridge and Manchester groups also contributed to a series of External Ballistics Department Reports (referenced E.B.D.) on ballistics based numerical and computational work. In one instance, E.B.D. 10, the report was written jointly.

Of the two Ministry of Supply differential analyser facilities, the Manchester SR(A) group was used for the widest variety of problems because of its attachment to the Headquarters department. In many ways it continued to operate as it had done before the war except that most of the investigations it carried out were war related and undertaken for government departments or research establishments. Problems were not carried out for their academic interest alone. The Ministry of Supply had recognized at the beginning of the war that the increased demand for ballistic data would present computational difficulties. By gaining access to the Manchester differential analyser and taking over the Cambridge Mathematical Laboratory, the Ministry of Supply went some way towards resolving some of the difficulties.

Scientists and mathematicians at other universities also contributed to the war effort by performing computational work and by acting as consultants to government departments. For example, before the war R. V. Southwell, Professor of Engineering at Oxford, had established a research group working on numerical methods to solve large numbers of simultaneous equations which arose from calculating the stresses in structures and frameworks. Southwell's technique was called the Relaxation Method and became widely used. Soon after war was declared in 1939 Southwell and his group began to perform work for the Ministry of Aircraft Production and others.

Some government departments had limited computing facilities of their own. For example, the Ministry of Supply External Ballistics Department establishment at Woolwich contained several National Accounting Machines, but they were not used to perform any outside work. Apart from the Ministry of Supply, only a few government departments tried to install extensive computing facilities to cope with the large amount of computation created by the war. In some research establishments solutions to the problems were put forward during the war but not taken up until after 1945. One example of such a case was at the Royal Aircraft Establishment (RAE) at Farnborough.

In the mid-1930s A. G. Pugsley[6] and R. A. Fairthorne of the Structural and Mechanical Engineering Department at the RAE had been trying to find efficient methods of solving large systems of linear simultaneous equations and calculating determinants which arose from their work on structural stress in aircraft bodies. Pugsley had installed desk machines at the RAE for this work and had also experimented with the Mallock simultaneous equation solver at Cambridge. In 1938 Pugsley and Fairthorne considered the possibility of using Hollerith punched card machines to perform the calculations and, over the following months, discussed the matter with Comrie, Sidney Chapman who used a Hollerith punched card installation at Imperial College for crystallography calculations, and the British Tabulating Machine Company.

As a result of his investigations, Pugsley presented a paper to the relevant sub-committee of the Aeronautical Research Committee (ARC) in February 1940 proposing the use of Hollerith machines for the calculation of determinants.[7] The report outlined a method for calculating determinants using punched cards machines and recommended that further investigation into such techniques be carried out. Pugsley's preliminary work was followed by two later reports presenting a more detailed discussion of the possibility of using punched card machines for scientific work (Frazer 1940; Schmidt and Fairthorne 1940). In October 1940 the committee chairman acknowledged the need for an improvement in computing facilities at the RAE but remained unconvinced that a Hollerith punched card installation within the Structures and Mechanical Engineering Department would be fully utilized and, hence, would be uneconomical. Due to the intense pressures of the war and particularly the national shortage of punched card equipment Pugsley abandoned this line of investigation.[8]

Admiralty establishments too were experiencing computational difficulties during the period 1939–1942. Unlike any other ministry the Admiralty already had a well-equipped computing centre in the NAO. The NAO, thanks to Comrie and his successor D. H. Sadler, was reasonably well equipped with a range of desk calculating machines but, perhaps more importantly, also had an experienced computing staff. Consequently the NAO received many requests from Admiralty research establishments and other service ministries to carry out different types of computation. Although NAO facilities were officially unavailable to outside personnel, Sadler took on as much additional war-related work as he could. By January 1942 30 per cent of the Office's time was spent performing special war-related computations such as preparation of ballistic tables for the Ordnance Board, Air Ministry, and Ministry of Aircraft Production. In addition astrograph tables and star curves for the astrograph were produced for the Air Ministry. The staff at the NAO also performed calculations of a more theoretical, less repetitive nature, such as mine design, for other Admiralty establishments.

Over and above the outside work which Sadler and his staff took on, the NAO's own work load increased during the war. One of the main causes of this was the computation of places of stars which would, under peacetime conditions, have been the joint responsibility of several international offices. To perform these calculations quickly and effectively Sadler proposed using a

Hollerith multiplying punch to compute the required tables. The Office did not have a Hollerith installation; in the past such equipment had been borrowed or hired when necessary. Although Sadler informed the Treasury in October 1940 of his intention to utilize a Hollerith multiplying punch for the task, the shortage of this type of equipment meant that BTM did not have a multiplying punch available at the time. Thus, after many delays, Sadler was given permission to engage the services of a BTM service bureau in April 1941.

The NAO did not, and could not, accept all the work which was brought to it during the early stages of the war. It had neither the staff nor facilities to do so. For example, in May 1941 the Ministry of Supply Assistant Director of Research approached Sadler concerning the preparation of a set of firing records for the External Ballistics Department of the Ordnance Board based at Woolwich. Sadler had to refuse the work on the grounds that his staff were already over stretched. The Office was already engaged in several large-scale jobs for the Ministry of Supply in addition to work for other Admiralty establishments and the Air Ministry. By July, however, two changes in circumstance had occurred and Sadler was able to contact the Assistant Director of Research, Ministry of Supply and accept the work. First, several jobs had been completed thus releasing staff and machines. Secondly, and more significantly, Sadler was able to state that

> Admiralty have recognized that there is a considerable amount of computational work required at the present time, and that this Office is ideally equipped to undertake it. They have accordingly sanctioned the appointment of an additional temporary assistant. (Sadler 8 July 1941)

Thus the use which other ministries and research establishments were making of the NAO facilities was officially acknowledged. The NAO was seen as a resource which should not, in wartime at least, be confined to the preparation of astronomical data but should be made more widely available. In effect the Admiralty had acknowledged that the NAO was acting as a limited computing centre to all the service ministries; something it had refused to do in the mid-1930s while Comrie had been Superintendent of the NAO.

THE FORMATION OF THE ADMIRALTY COMPUTING SERVICE

Recognition by the Admiralty that the NAO was becoming heavily involved with additional military computations was, in itself, not enough to prompt the creation of a formal computing service. Sadler, because of his continual involvement with individual computations, was aware that the problem of insufficient computing facilities was not confined to the Admiralty alone but to the whole of the scientific civil service and the other service ministries. Consequently, in 1941, he proposed that a central computing facility be set up for use by all government research establishments and the armed forces. His proposal came to nothing because recognition of the problem was not yet widespread. The less ambitious suggestion to centralize computation within the Admiralty came as a result of other Admiralty personnel reaching the same conclusions as Sadler.

In 1942 J. A. Carroll, Professor of Natural Philosophy at Aberdeen University and well-known astronomer and physicist, joined the Scientific Research and Experiment Department (SRE) of the Admiralty as Assistant Director of Scientific Research. The prime function of the SRE Department was to administer naval research establishments from Whitehall and serve as a means of communication and liaison between them. It also had the task of keeping Admiralty establishments informed of scientific developments taking place outside the ministry and of employing external consultants when required. In the autumn of 1942 John Todd[9] also became attached to the SRE Department. At the beginning of the war Todd, a mathematician, had been assigned first to a degaussing range and then to the Mine Department of HMS Vernon. At the former Todd came to the conclusion that he was not an applied mathematician nor an experimental physicist, and at the second found himself employed doing minor calculations concerning electric circuits. Todd recalls

> Observing this and similar matters I realized that pure mathematicians, such as I, could be more useful in dealing with computational matters and relieve those with applied training and interests from what they considered as chores. (Todd 1983)

Before joining the Admiralty, Carroll had developed an interest in table-making and had consulted Comrie on the merits of desk calculators for such work. Soon after his appointment to the SRE Department, Carroll became aware that a significant amount of time was being spent by experimental staff performing calculations. In some cases, for example where the calculations were inseparable from their experimental background or where they were brief, this was not important. Carroll, however, concluded that scientific effort was being wasted in performing long, repetitive calculations particularly in cases where the specialized machinery or knowledge needed to perform the work efficiently was not available. With the wartime pressure for rapid research results Carroll felt that it was important that computation be handled quickly and effectively by *specialists*.

As a result of these feelings within the SRE Department Sadler was asked, in late 1942, to investigate the computational requirements of the numerous Admiralty research establishments. Sadler's report to the Director of Scientific Research (DSR) summed up the situation. It stated:

> (a) Some of the computation arising in the establishments staffed by D.S.R. can, with advantage, be centralised.
>
> (b) Research and experiment give rise to a proportion of work unsuitable for centralisation so that there will always be a need for local facilities and, in some cases, specialised local staff.

(c) Centralisation would not only lead to acceleration in the presentation of results but would make possible theoretical research involving calculations too lengthy to be tackled locally (Carroll, 1943, p.1)

Sadler anticipated that a central Admiralty computing centre would undertake long, routine calculations, calculations on behalf of Admiralty research establishments for which equipment or expertise did not locally exist, mathematically complex work, and work which would, under normal circumstances, not have been undertaken at all.

This report confirmed the opinions of Carroll and Todd. Once the decision had been made to set up an Admiralty Computing Service (the ACS) it was quickly implemented. Permission from the Astronomer Royal regarding the use of NAO staff and facilities was sought and granted. In early 1943 the ACS, based at the NAO, was created. Additional mathematical and computing staff were attached to the NAO to carry out the work of the service. They were trained by Sadler. In order to use the service Admiralty establishments had to approach a small administrative staff at the SRE Department in Whitehall which then passed the work on to the NAO under Todd's supervision. The ACS, therefore, operated on two levels: it was administered from Whitehall and the actual computational work was carried out at the NAO.

There were three reasons for this two-tier administrative system. First, issues of priority would natural arise where several establishments were competing for resources. It was felt that questions of priority would be more smoothly overcome by the SRE Department rather than the NAO as the former was an already recognized line of communication between Admiralty establishments. Secondly, it was decided that the staff of the ACS should include a number of experts from various mathematical, statistical, and computational fields. This was to be achieved by employing mathematicians and scientists working in other Admiralty departments, other service ministries, and universities as consultants to the ACS. The SRE Department already employed outside specialists from a variety of fields and thus, again, it was the obvious administrative procedure. Thirdly, it was proposed that the SRE Department act as a clearing house for all the problems brought into the ACS. Todd's staff divided the work sent to them into two classes: mathematical and computational. Those problems which were mathematically complex and which did not raise any questions of security could be quickly passed to the appropriate consultants. Once the mathematical problems had been solved the individual establishments were asked to make the decision locally as to whether the work could be carried out on site or whether it should be performed at the NAO. If the latter course was chosen then the SRE Department handled the necessary arrangments.

The role of the SRE Department was to liaise between the individual Admiralty establishments and the NAO. In cases where the NAO did not have the necessary machinery to carry out the work, that is, if Hollerith machines or large analogue devices were required, the SRE department was to make arrangements for the work to be carried out elsewhere. The main disadvantage of the system was that Sadler and his staff at the NAO had very little contact with the Admiralty establishments in which the work originated. This meant that the NAO often had no background knowledge of the physics behind a calculation and had no control over the way in which data was recorded or presented. The way in which experimental data is presented can have a critical effect on the accuracy or type of analysis which can be performed. In many cases if consideration is given before the experimental data is collected to the type of results required it may be possible to minimize the amount of manipulation needed to get the data into a convenient form for analysis. On the other hand, the two-tier administrative system did relieve the already hard-pressed NAO from the burden of carrying out the considerable amount of administration which was required in running such a service.

Thus, by March 1943, the ACS had been set up and was functioning as a computing service for Admiralty establishments. Todd visited all the Admiralty research establishments which came under SRE Department jurisdiction to inform them that the ACS had been set up. Sadler too visited some establishments to advise on computing machinery to be installed there. Because few people were providing this type of computing service the ACS had some difficulty obtaining experienced computing staff and consequently the ACS began on a limited scale. Those whom Sadler did manage to recruit included L. Fox who had been a student of R. V. Southwell at Oxford, E. T. Goodwin a member of Lennard-Jones's exterior ballistics team at Cambridge, F. W. J. Olver, K. Blunt, and H. H. Robertson.[10] From the annual reports of the Astronomer Royal it is obvious that a large proportion of the work carried out by the NAO during 1943–5 was for the ACS and carried out at the expense of astronomical and navigational computations.

Details of individual computations carried out by the ACS have not been preserved but it is clear that a wide range of problems was dealt with including the theory of supersonic flow, trajectory work, stresses in turbines, and a statistical investigation into the night vision capabilities of naval personnel. Hence it can be concluded that many of the Admiralty research establishments to which the ACS was available were making use of the service. In 1946 Sadler and Todd stated that by the end of the war

> the Admiralty Computing Service was providing a fairly comprehensive mathematical and computational service which was not only meeting all the demands from Admiralty sources but was also able to offer informal assistance to the other Services, Government departments and contractor who had no comparable facilities at their disposal. (Sadler and Todd 1946, p.571)

But how far did the ACS go towards a truly comprehensive computing facility?

THE ACS AS A COMPUTING CENTRE

Intuitively it is easy to say that the Admiralty Computing Service was a computing centre. It was set up specifically to centralize computation within the Admiralty. However it is still useful to consider the ACS in terms of the nine characteristics identified in Chapter 1 as describing a computing centre. Namely users, personnel, machinery, computing activity, advisory role, numerical research, computing machinery research, publications, and library facilities.

Admiralty research establishments were the only official users of the ACS, but informal advice was given to workers outside the Admiralty when requested. All users had first to approach Todd at the SRE Department before their problems were passed on either to a consultant or to the NAO. In this respect the two-tier administrative structure was a cumbersome one for users as they never had direct contact with computing staff.

All the routine work performed by the ACS and the administration of the computations were carried out by NAO personnel many of whom had been trained in numerical computation by Sadler. For more complex work the Admiralty employed a number of consultants. From the outset, however, the ACS suffered from a shortage of trained staff. It is difficult to determine how many computing staff were employed specifically by the ACS as they were well integrated with the staff of the NAO and supervised by Sadler, but at any one time there were approximately eight. This shortage of experienced staff was due, in part, to conscription but was aggravated by the lack of training in practical computing techniques given in British universities. This lack of trained computing staff caused considerable concern and was to result in a number of articles, written after the war, calling for such courses to be introduced.[11]

The list of consultants who offered their services to the ACS is impressive. They include N. Aronszajn, W. G. Bickley, L. J. Comrie, E. T. Copson, J. Cossar, A. Erdélyi, P. P. Ewald, H. Kober, J. Marshall, J. C. P. Miller, E. H. Neville, and H. L. Seal and S. Vajda from the Admiralty statistical section. Thus, the ACS reaped some of the benefits of a large staff, experienced in a wide range of fields without the cost of their being full-time employees of the ACS. The ACS was not large enough to justify a large highly qualified staff on a full-time basis, nor would it have had the income to do so.

The type of machinery used to perform ACS computations was limited to that already available at the NAO, that is, desk calculators and National Account Machines. A comprehensive computing service of the early forties might have been expected to possess a differential analyser, a Hollerith installation, and perhaps several more specialized pieces of equipment. However, the ACS was not sufficiently large to justify the expense of either a differential analyser or a Hollerith installation. In any case it is unlikely that machines of this type could have been obtained during the war. Sadler and Todd (1946) reported that, between 1943 and 1946, the need for a differential analyser never arose and that a Hollerith installation would only have been used, on average, for two months of the year. Todd had considered installing Hollerith equipment and was caught between the two pressing factors of computing efficiency and financial economy. He resolved the problem by making arrangements to make use of machines located elsewhere when the need and opportunity arose.

The type of work performed by the ACS fell into two main classes: large, repetitive calculations and those involving complex mathematics. The advisory role the ACS played was that of an organization which Admiralty establishments could consult on computational techniques and on the use of the machinery at its disposal. By 1947 the ACS had performed over 100 investigations besides dealing with enquiries which required only advisory assistance.

In cases where mathematical or numerical research was required for the solution of a particular problem the ACS consulted experts in the appropriate field and developed the ideas which they supplied. The SRE/ACS reports were often the results of these investigations. The ACS did, therefore, carry out some degree of research and development into numerical methods but only as a result of a specific need. In most cases they were carried out on the behalf of ACS users, but reports such as the Dictionary of Laplace Transforms (SRE/ACS nos. 53, 68, 71, 102, and 108) illustrate that mathematicians, such as Cossar and Erdélyi, were using the ACS to develop methods and provide computing tools for the scientific community at large.

The ACS was not equipped, nor was it set up, to undertake research into the design and subsequent development of calculating machinery. It did, however, offer information to Admiralty establishments on a wide range of computing equipment, from desk calculators and Nationals to punched card machines, differential analysers, and, after the war, the ENIAC. The one exception to this was the design of an electromechanical device by A. Baxter, from the SRE Department in Whitehall, which calculated the root and mean square errors made by trainee range-finder operators. This was however not a usual part of the work performed by the ACS.

The initial objective of the ACS was 'to obtain the numerical results required in problems of war research and to make those results available to the particular establishment concerned as early as possible (Sadler and Todd 1947, p.290). The preparation of reports and publication of results was therefore of less immediate importance. At first the few reports which were prepared were distributed by the NAO (usually in the form of duplicated manuscript, or occasionally typescript, copies). However, by 1944 the publication of a series of monographs describing pure and applied mathematical techniques had begun to be distributed to Admiralty establishments administered by the SRE Department. The purpose of these monographs was to familiarize scientific workers with these techniques so that they could either deal locally with computations which were too small to be efficiently centralized or recognize when a problem could be solved by a standard method. The SRE/ACS report series also included mathematical tables which were either unpublished, or not published in a convenient form. Appendix 2 gives a selection of the titles in the SRE/ACS series and serves to illustrate the variety of work performed by the ACS. Many of the SRE/ACS reports are anonymous but others were prepared by the consultants associated with the ACS. After the war these SRE/ACS reports were reviewed on several occasions in *Mathematical tables and other aids to computation* where their contents were of interest to its readers. An article describing the ACS itself was also featured in 1947 (Sadler and Todd). *Nature* also published an item concerning the ACS (Sadler and Todd 1946). The ACS, therefore, not only published reports of its work but also, after the war, made the public aware of the purpose of its work.

It is unlikely that the ACS would have had a library of its own but would have made use of the existing collections at the NAO and at the SRE department. The department of SRE in Whitehall maintained the Admiralty Technical Records Section for the use of Admiralty establishments. A set of the SRE/ACS monographs would have been held here and thus were available to all users. It would hence have been superfluous to create an additional library for the ACS.

There are no direct records available to determine whether individual Admiralty establishments found the ACS useful. Sadler and Todd (1946) and Conolly (1947), who later took over the administration of the ACS, have stated that the ACS was successful in easing problems in the Admiralty within the limits of its terms of reference. Thus the ACS succeeded in fulfilling the aims of its administrators and creators. It was, therefore, all it needed to be.

Despite the partial, and in some cases total, fulfilment of the nine criteria of a computing centre listed above, the ACS cannot be described as a comprehensive computing centre for three reasons. First, its use was officially restricted to Admiralty personnel; secondly it had a limited range of calculating machinery in its possession and access to larger devices elsewhere was by no means automatically available; and thirdly it did not undertake research into new machines. In addition to these limitations the ACS also did not fulfil the entire concept of a computing centre in another important respect. It was divided, both physically and conceptually, between the NAO at Bath and the SRE in London. Therefore, although the ACS was a practical solution to problems which existed at the time, it did not totally embody the centralized computing principle. It was, however, an extremely important and visible step towards that goal.

ENDNOTES FOR CHAPTER 6

1 Most of the details known about the war work of the SCS have been extracted from SCS information pamphlets (SCS 1938*b*, 1946, 1950*a*), Comrie's own reminiscences (Comrie 1948*b*), and from publications describing work to which the SCS contributed (Comrie 1942; Hodgkin 1949; Jones 1978).

2 Predictors were analogue instruments used to direct the fire of anti-aircraft guns.

3 Kine Theodolites were used to photograph objects from known positions. By inspection of the developed photograph through an evaluator (an instrument providing illumination, magnification, and scaling data) the displacement of the photographed object could be determined. Using Kine Theodolite readings on the same object photographed two or three miles apart the height and bearing of the object could be calculated using triangular geometry. To use this type of data effectively it was essential that the results were calculated as quickly as possible. Comrie found that the Single Marchant was the most suitable machine for the task.

4 A conclusion also reached by the Aberdeen Proving Ground in the United States (Ministry of Supply n.d.).

5 During the war the term fire control equipment described the analogue control mechanisms for targeting anti-aircraft guns and similar weapons.

6 A. G. Pugsley, later Sir Alfred, was then head of the Structural and Mechanical Engineering Department of the RAE. In 1945 he was appointed Professor of Civil Engineering at Merchant Venture College, Bristol. Pugsley was chairman of the Aeronautical Research Committee 1952–7.

7 The ARC was an advisory committee which directed aeronautical research in the Air Ministry, the Ministry of Aircraft Production (created in 1940), and the NPL Aeronautical Division. Although influential it had no executive powers nor any financial control over the establishments it advised.

8 During the war punched card equipment was heavily used by the armed services for administration and logistics. The Hollerith and Powers factories too were partially given over to the manufacture of supplies for the armed services. Thus a shortage of punched card equipment developed during the course of the war.

9 John Todd later joined the United States National Bureau of Standards Institute for Numerical Analysis before joining the California Institute of Technology.

10 Fox, Goodwin, Olver, Blunt and Robertson all subsequently went on to join the NPL Mathematics Division. Fox, Goodwin, and Olver later made significant contributions to the emerging field of numerical analysis.

11 For example see Erdélyi and Todd 1946 and Sadler and Todd 1946.

7

The creation of a national computing centre

THE DEPARTMENT OF SCIENTIFIC AND INDUSTRIAL RESEARCH INITIATIVE

Within a year of the Admiralty Computing Service becoming operational Todd, Sadler, and A. Erdélyi, an Admiralty scientific consultant during the war, realized that the degree of centralized computation they could offer within the Admiralty did not go far enough. The ACS did not operate on a large enough scale to run a fully equipped computing service. It was not economically feasible to install punched card machines or a differential analyser for example. In some cases, therefore, computing efficiency was sacrificed because the most suitable type of equipment was not available to the ACS. Those responsible for the ACS, Todd and Sadler, were of the opinion that only centralization at the national level would achieve the economies of scale necessary to provide a fully equipped service.

In response to the opinions expressed by Todd and Sadler, Carroll approached Sir Edward Appleton, secretary of the Department of Scientific and Industrial Research (DSIR), proposing that a national computing organization be set up. This proposal was accompanied by a supporting document probably written by Todd, Sadler, and Erdélyi, *Memorandum on the Centralization of Computation in a National Mathematical Laboratory* (Anon. n.d.). From this document it is clear that the motivation behind the proposal was the obvious need within the other service ministries for a facility similar to the ACS. The memorandum was based on the NAO's experience of performing computations on behalf of several government departments, the experience of the ACS, the work of Comrie at the SCS, and the existence of the New York Mathematical Tables Project.[1]

The main case for centralization presented by the Admiralty memorandum was the same as that presented by Sadler for the ACS in 1942:

(1) efficiency of computing methods by trained staff and specialists with access to a wide range of calculating machines; and
(2) economy of machines and staff.

However, the experience of running the ACS on this basis had identified four additional arguments in favour of centralization. These were:

(1) the benefit of accumulated computing experience by a regular staff which could efficiently deal with the recurrence of similar problems which may stem from a variety of sources;
(2) the ability of trained staff to identify, and hence carry out, research into areas where new numerical methods and machines were most needed;
(3) the benefit of having a central, specialized library of books, tables, reports, and papers on occupational and allied subjects;
(4) a central administrative structure to act as a liaison between establishments and provide for the organized and systematic dissemination of relevant information.

It was, therefore, already recognized that the role of a National Mathematical Laboratory extended beyond running a service bureau to which the user presented a problem and received a solution in return. Such an establishment, it was proposed, should also carry out research into both machines and numerical techniques and act as a clearing house to hold and pass on information. In conclusion the Admiralty stated several features which it saw as desirable in a National Mathematical Laboratory and outlined how such a laboratory should be run.

The first feature suggested in the memorandum was that any National Mathematical Laboratory should 'Act generally in the sphere of (numerical) mathematics and computation in a way similar to that in which the National Physical Laboratory acts in the sphere of physics' (p.3). The proposed National Mathematical Laboratory should not only take on computational work for government departments but should also extend its facilities to industry and the universities. In addition the laboratory should carry out fundamental research into numerical methods and machines. It was intended that the laboratory should become the leading authority on computational methods in Britain.

Secondly, the memorandum advised that the National Mathematical Laboratory be non-departmental and should, therefore, come under Department of Scientific and Industrial Research administration. This followed naturally from the two tier administration of the ACS and the proven advantages of having a central body which could objectively discuss priorities.

The memorandum next outlined how the Admiralty envisaged a comprehensive computing centre should be organized. It was proposed that there be five individual sections each devoted to different types of computational work. The memorandum stated that the responsibilities of the different sections should be:

(1) mathematics, applied to numerical computation;

(2) the provision of a comprehensive collection of mathematical tables, and the computation of such new tables as may from time to time be required;
(3) general computation by machines including Hollerith, National and other installations, and the development of methods and machines;
(4) general computation by instruments such as differential analysers, cinema integraphs, Fourier transformers, Mallock machines, etc., possibly including graphical and nomographical methods and the development of such instruments to meet requirements;
(5) statistical mathematics, tables, methods, and computation (but not the collection of data).

The remainder of the memorandum concentrated on the three ways in which the Admiralty saw the National Mathematical Laboratory being used. First, any such laboratory should carry out all types of computational work for any government department which needed assistance and make computational facilities available to those who chose to perform their own work but did not have the necessary equipment. Secondly, it should advise all government departments on the installation of computing facilities and liaise with the Treasury over such matters. Thirdly, a mathematical laboratory should liaise closely with other institutions such as university mathematical laboratories, the British Association for the Advancement of Science Mathematical Tables Committee and overseas computing centres. Any new developments in computing techniques could therefore be made available to government establishments through the laboratory.

The most significant departure from the original ACS concept is demonstrated by the title 'National Mathematical Laboratory'. The Admiralty, more specifically Todd, Sadler, and Erdélyi, were proposing more than a large-scale version of the ACS. They were presenting the case for a national effort to be made to develop numerical mathematics and computing as a specialized subject.[2] The ACS was never intended to fulfil this more general role; it was created to quickly supply additional computing power necessary under wartime conditions. The proposed National Mathematical Laboratory was intended to be a permanent organization.

Unlike the proposal which Sadler had made in 1941, this approach to the DSIR in 1944 bore fruit and moves were made to consider the question further. There were three reasons why this later proposal was considered more seriously. FIrstly, this proposal made to the DSIR had the approval of the Director of Scientific Research (Admiralty) whose high rank led to its prompt consideration by the DSIR. Secondly, the proposal was based on the experience of the ACS and not just the work of the NAO; the ACS had successfully demonstrated the need for such a service. Thirdly, and perhaps most importantly, the question of establishing a National Mathematical Laboratory had already been voiced in other quarters.

One such voice was that of Sir Charles G. Darwin,[3], Director of the National Physical Laboratory (NPL). In March 1943 Darwin informally expressed to the DSIR Advisory Council the opinion that a Mathematical Department should be set up at the NPL to carry out the computations arising within government establishments. He was, in essence, calling for the establishment of a national computing centre. The possibility of a National Mathematial Laboratory was also being freely discussed in a variety of circles and Darwin had sounded out the views of his contemporaries. For example the establishment of such a Laboratory was discussed by Lennard-Jones, Darwin, and R. H. Fowler[4] during May 1943 (Lennard-Jones 1943) and the possibility of the Cambridge Mathematical Laboratory taking on the prescribed role mooted. From this conversation it is evident that Hartree too had been consulted on the issue.

Thus, before the Admiralty proposal had been submitted, the ground work had already been done in bringing the problem to the attention of influential scientists. As a result, the proposal from the Admiralty was considered and the DSIR contacted various ministries which were thought to have an interest in the work such a laboratory might perform. On the results of this preliminary survey it was agreed that the proposal to set up a National Mathematical Laboratory be further investigated by an Interdepartmental Technical Committee to 'examine the question of whether it is desirable to establish under government control a Central Mathematical Station, and, if so, to make recommendations on the form it should take' (DSIR 1944, p.1).

THE 1944 REPORT OF THE DSIR INTERDEPARTMENTAL TECHNICAL COMMITTEE

The Interdepartmental Technical Committee set up by the DSIR to consider the question of a Central Mathematical Station drew its 20 members from 11 different government departments. It was chaired by Darwin in his dual role as Director of the NPL and Chief Scientific Advisor to the War Office. Table 1 lists the committee members. The number of representatives from each department illustrates their relative importance as potential users of centralized computing activities. Four of the 20 committee members were taken from the Admiralty; Sir Harold Spencer-Jones, the Astronomer Royal, Carroll from the SRE Department, Sadler from the NAO, and Todd as ACS organizer. Of these Carroll, Todd, and Sadler had been partly responsible for bringing the question to the attention of the DSIR.

The Ministry of Supply was also well represented providing three members of the committee: Hartree, Maccoll, and Womersley. Hartree was an acknowledged computing expert and was responsible for the Manchester differential analyser project.[5] Maccoll represented the Armament Research Department under Lennard-Jones and was involved with the external ballistics work performed at the Cambridge Mathematical Laboratory. Womersley a

Table 1. DSIR Interdepartmental Technical Committee membership

Sir Charles Darwin (Chairman)

Member	Representing
Sir Harold Spencer-Jones	Admiralty
Prof. J. A. Carroll	Admiralty
Mr D. H. Sadler	Admiralty
Mr J. Todd	Admiralty
Dr F. Yates	Agricultural Research Council
Prof. W. J. Duncan	Ministry of Aircraft Production
Dr S. H. Hollingdale	Ministry of Aircraft Production
Mr J. R. N. Stone	War Office Central Statistical Office
Mr A. W. Taylor	Customs and Excise
Dr Christopherson	Ministry of Home Security
Dr David	Ministry of Home Security
Prof. D. R. Hartree	Ministry of Supply
Dr J. W. Maccoll	Ministry of Supply
Mr J. R. Womersley	Ministry of Supply
Major Gen. G. Cheetham	Ordnance Survey Department
Mr A. W. Mattocks	Treasury
Mr G. F. Peaker	Treasury
Major E. H. Thompson	War Office, Directorate of Military Survey
Dr S. Goldstein	DSIR

statistician and the third Ministry of Supply representative, was at that time head of the Ministry of Supply Advisory Service on Statistical Methods. Womersley had worked with Hartree before the war while employed at the Shirley Institute of the British Cotton Industry Research Association. Womersley had also worked for the Armament Research Department at Woolwich where he had been responsible for the construction of a model differential analyser.

The Ministry of Aircraft Production (MAP) had two representatives, Duncan and Hollingdale. Duncan was a member of the Aeronautical Research Council, in particular he had sat on the Oscillation Sub-Committee to which Pugsley had presented the case for the installation of Hollerith machines at the RAE in 1940.[6] During the war Duncan was also involved in a wide range of aeronautical and defence projects at the RAE and at the Air Defence Department at Exeter. He was thus well informed regarding the computational requirements of the MAP.

With the exception of the two Treasury representatives and Cheetham from the Ordnance Survey Department, the remainder of the committee represented departments with primarily statistical interests.

The first task undertaken by the committee was a review of the computational and mathematical work being done by government departments. The committee was also asked to predict the expected post-war needs and assess what would be made of a Central Mathematical Station. This survey included all the departments represented on the committee, except the Treasury, plus the Medical Research Council and the Ministry of Works. This review, given in Appendix A of the report, did not convey a universally enthusiastic welcome for the proposal and, in some cases, departments clearly stated that they would only make very limited use of such a service.

The only department consulted which declared that it would not use a Central Mathematical Station was the Ministry of Home Security. During the war the Ministry of Home Security was engaged in mathematical work concerned with civil defence and operational research. It was unlikely that this work would continue after the war and hence the Ministry could not foresee that it would have a need for the station. HM Customs and Excise also stated that it would have only a limited use for such a station as it performed very little mathematical work. Its main computing work involved large-scale data processing. It was willing, however, to allow access to their punched card installation during idle periods, if there was a central administration to handle the arrangements.[7]

Only the Admiralty and the War Office anticipated using the station on a regular basis. The Admiralty foresaw that the station would take over the computational work of the ACS and relieve the NAO of its ACS responsibilities. The War Office suggested that its Directorate of Military Survey would continue to require the preparation of geodetic tables and the computation of triangulations after the war and would, therefore, make use of such a station. During the war the SCS carried out a considerable amount of this type of work and the War Office stated that this could be taken over by the Central Mathematical Station.

Three other departments considered using the station as a computing bureau: the Ministry of Aircraft Production, the Medical Research Council and the DSIR. All of them stated clearly that they could use the facilities of such a station for a small part of their work or on occasions when they had too much work to handle locally. Indeed, in every case considered, with the exception of the Ministry of Home Security, the individual department consulted wanted to maintain its own computing staff and equipment.

Four departments, the Admiralty, the Agricultural Research Council, the Ministry of Supply, and the War Cabinet Central Statistical Office welcomed the proposed station as an avenue through which they could channel the complex and specialized problems which occasionally arose. In a similar way the Ordnance Survey Department of the War Office, the Ministry of Works and the Admiralty felt that the existence of a centre to which they could go for advice would be an important feature. Of the 12 departments reviewed, only the Admiralty, the Agricultural Research Council, the War Cabinet Central Statistical Office, and the War Office conveyed any expression of enthusiasm for the establishment of such a station.

As a conclusion to the review of computing and mathematical activities prepared for the committee's report, a brief mention was made about the work being done outside the scientific civil service. The extent of Hartree's work on the differential analyser at Manchester University and the uncertain future of the Cambridge Mathematical Laboratory were put before the committee. The most important comment in this section was made under the headings 'Scientific Computing Service Ltd' and added further weight to the argument that a Central Mathematical Station was necessary. It was reported that the

> Scientific Computing Service Limited., accepts contracts to carry out computations. This it had done over a wide variety of subjects. The successful establishment of such a business and its expansion shows that there has been a real need for it. (DSIR 1944, Appendix A, p.3)

The committee was presented with the very diverse needs of a set of departments of which almost one half were concerned with statistical and not mathematical work. In reply the committee categorized the functions of a Central Mathematical Station into two classes: general mathematical computations and statistics. In both areas the committee considered that the most important role of a Central Mathematical Station would be the existence of a body of experts which would not only undertake work on special problems but would also develop new computing machines and techniques. The computing service which such a station would provide was seen as fundamental to the station's role and as a source of income to help fund the research activities of the station.

The conclusions of the DSIR Interdepartmental Committee were presented to the DSIR Advisory Council on 10 May 1944 (DSIR 1944). The final report recommended that a Central Mathematical Station be set up and that it should

> —deal with problems arising out of statistical science, especially in the principles of its application to industrial research, development and production, including users' requirements, but not to handle detailed statistical problems of a descriptive type such as arise in economics, sociology and biology, though it would be ready to act as consultant on the mathematical aspects if required;
>
> —provide computing services for Government Departments, industry and Universities. In cases where the work of a Department is continuous, voluminous, uniform or specialized in character, such Department could employ a whole time staff of its own;
>
> —act as consultant for Government Departments, industry and Universities in the use of mathematical and statistical techniques. The centre would undertake work

> in cases where a Department needs it only at irregular intervals. It could also help in carrying overloads, and in arranging for any one Department which was incompletely occupied for a time to help in carrying the overload of another.

However, the committee took the suggested role of the proposed Mathematical Station further than these three recommendations. The committee perceived

> —research into new computing methods, including the design of new instruments

As the most important function of the station. This new emphasis on machine research was a result of the interests and experiences of the individual committee members all of whom appreciated the need for continuous research.

The remaining two functions outlined in the committee's list of six were:

> —to act as a liaison between British workers in the field, to collaborate with the BAASMTC and the University Mathematical Laboratories, and to disseminate information on computing machines and methods by the distribution of reports and the establishment of a library of mathematical tables;
>
> —to consider the need for mathematical tables and, if necessary, compute them.

It is interesting to compare this list with the functions of a proposed National Mathematical Laboratory put forward by the Admiralty memorandum to the DSIR. All of the Admiralty proposals were contained in the six functions listed by the DSIR Committee; but the committee put a much greater emphasis on research and development of new machines as a critical part of the station's work. To the Admiralty the concept of a computational and advisory service was of greater importance than the development of new machines and methods, although this was by no means rejected as a function of such a laboratory. The consultation process and the committee membership, had, therefore, changed the proposed National Mathematical Laboratory from an enlarged ACS-type organization into an institution which was to perform basic research into numerical mathematics and computing machinery in addition to acting as an advisory centre and computing bureau.

Having established a need for a Central Mathematical Station and set down its primary functions, the committee turned its attention to the practical requirements and location of such a laboratory. The conclusions reached were based on the experience of the NAO, the work of the differential analysers at Manchester and Cambridge, and the role played by the Ministry of Supply Central Statistical Office set up in 1939 to prepare production forecasts. It was proposed that the station be provided with a staff of 25 scientific officers and about 50 ancillary staff. Table 2 illustrates how the staff were to be distributed. The cost of such a staff was estimated at approximately £34 000 per annum. The station was to be equipped with a variety of desk machines, adding and

Table 2. Breakdown of estimated staff requirement for the proposed Central Mathematical Station

	Scientific officers	Ancillary staff
General computation and research	12	24
Instruments e.g. differential analyser	3	6
Statistics	7	14
Expected increase for new machines	3	6
Total estimate	25	50

listing accounting machines, a Hollerith installation, a differential analyser of larger capacity than any other in Britain, a Mallock Machine, and a few minor special machines.[8] The cost of the machines was estimated at under £50 000 at 1938 prices plus between £2 500 and £4 000 per annum rental for the Hollerith equipment. It was also decided to locate the Mathematical Station at Teddington, near the NPL, on the grounds of close proximity to an established intellectual centre, accessibility to government departments, and on-site workshop facilities. The benefits of an established administrative structure and direct access to workshops resulted in the decision to set up the Central Mathematical Station as a division of the NPL.

On receiving and considering this report the DSIR Advisory Council recommended that 'the proposed scheme for the establishment of a Central Mathematical Station be proceeded with, subject to consideration by the Chairman of any observations or recommendations made by the Executive Committee of the NPL (DSIR 1942–7, AC Minute 88 1943–4, 10 May 1944). By 12 July 1944 the NPL Executive Committee had approved the proposal and by October Treasury approval had been granted.

THE NEW NPL MATHEMATICS DIVISION: 1945

Even before Treasury approval for the establishment of a national computing centre at the NPL had been officially granted, the executive committee of the NPL had begun the process of appointing a Superintendent. As early as June 1944 a sub-committee had been set up to interview and select suitable candidates. The sub-committee was almost entirely made up from members of the executive committee. The members were D. Brunt, FRS; R. H. Fowler, FRS (for reasons of ill health later replaced by S. Chapman); D. R. Hartree, FRS; A. V. Hill, FRS, chairman of the NPL Executive Committee; and J. Lennard-Jones, FRS. The Ministry of Labour and National Insurance was also represented on the sub-committee. Hartree was the only member of the sub-committee to have been a member of the 1944 DSIR Interdepartmental Committee which recommended the establishment of a National Computing Centre at the NPL.

The sub-committee interviewed only two candidates, D. H. Sadler and J. R. Womersley, both of whom had sat on the DSIR Interdepartmental Committee. Sadler, Superintendent of the Nautical Almanac Office, had been invited to apply for the post by Darwin, Director of the NPL, but had been unenthusiastic about taking up such an appointment (Sadler 1984*a*). Womersley was, at that time, working in S.R. 17, the Ministry of Supply Advisory Service on Statistical Methods. By September 1944 the sub-committee had reported and the Superintendentship of the Mathematics Division had been offered to Womersley.

From late 1944, therefore, Womersley had the task of building up the NPL Mathematics Division from the terms of reference given in the Interdepartmental Report. To supplement the general outline of what the division's activities were to be, the Interdepartmental Technical Committee provided an initial research programme. This programme (given in Appendix 3A and summarized in Appendix 3B) outlined 12 areas of research which the committee felt should be the concern of the division. In 4 areas (items 3, 5, 7 and 12 in Appendix 3A) the committee recommended the construction of specialized machines to perform specific types of calculations such as solving integral equations or evaluating inverse Laplace transforms. The closest the programme came to recommending general research on computing machine technology was the call for an investigation into the possible application of telephone equipment to computing machinery (Appendix 3A, item 2). Five of the 12 items listed in the research programme were directly related to statistics, hence representing fairly the large number of statistical users which the DSIR had consulted. In addition to the research programme presented in the Interdepartmental Committee's Report, the NPL Research Programme for 1945–6 included the 'Development of electronic counting devices for rapid computing' (NPL 1942–7, Minute 4228, Report E851, October 1944).

In the light of these guidelines, and his own experience at the Ministry of Supply and the War Office, Womersley presented his plans for the Mathematics Division to the NPL Executive Committee in December 1944. Womersley proposed to organize the division into three separate sections to deal with different aspects of the Division's work. The sections were to deal specifically with:

(i) computing on commercial machines,
(ii) analytical engines and computing development, and
(iii) statistical methods.

The first of these three sections, computing on commercial machines, was to undertake the non-statistical computing bureau and consultancy functions of the division. It was proposed that it should take over the role already played by the Admiralty Computing Service and any numerical research being carried out under its name. Womersley went so far as to suggest that staff from the ACS be transferred directly to the NPL. No attempt was made, at this point, to elaborate on the types of machine with which this section was to be equipped. It was also suggested that this section build up a specialist library which was to be more extensive than either the SCS or Admiralty collections and which would include all the known published mathematical tables.

Womersley proposed that the major task of the section devoted to analytical engines and computing machinery development would be the construction of a large differential analyser initially having 18 integrators but with provision for expansion to 24. Womersley also suggested the development of a 'production model' differential analyser. This device was to be considerably smaller than the large machine proposed and was to be suitable for laboratory use. The possibility of constructing other analogue machines to evaluate integrals, solve simultaneous equations, and calculate the roots of complex polynomials was also put forward.

Womersley was aware, however, of the economic and philosophical conflicts between the construction of specialist machines and the development of efficient desk machine techniques to carry out the same type of calculation. He discussed with the executive committee of the NPL the importance of carrying out research into the possibility of using electronics, particularly automatic telephone equipment, as general purpose computing devices. He anticipated that to carry out research into machines using electronic components the NPL would need to contract out the construction of electronic components. Womersley stated that he was soon to take a trip to the United States on behalf of the Ministry of Supply and would take the opportunity to 'undertake a comprehensive tour of universities and Government and commercial organizations interested in the field covered by his plans' (NPL 1942–7, 19 December 1944, p.6). It was therefore anticipated that the type of machinery developed by the Mathematics Division might be influenced by events taking place in the United States.

Womersley had also made plans for the third Mathematics Division Section, Statistics. Because of his current position within the Ministry of Supply, Womersley had reached agreement with the Ministry's statistical group (S.R. 17) to limit the group's activities to production problems for the armed services. All other statistical work covering ballistics, quality control, etc., but not operational research, was to be channelled to the NPL.

Womersley's initial proposal concerning the work of the Mathematics Division made few references to the type of machinery to be installed, the number of staff to be taken on, or any potential users of the Division's facilities. Rather the proposal clearly stated that the Division was going to provide a computing service, play a consultancy role, undertake numerical research, investigate and construct new types of computing machinery, and maintain a specialist library.

Once Womersley had established these broad, long-term plans for the NPL Mathematics Division, he gave the executive committee a brief outline of how he intended to put these plans into operation during 1945. He stated that much time during 1945 would be devoted to recruitment and training of staff and to performing computations for other NPL divisions. He anticipated that very little work would be done towards the construction of the new differential analyser. In the event recruitment to the Mathematics Division was delayed because of restrictions imposed by the Ministry of Labour and National Insurance governing the appointment of staff before the end of the war had been officially declared. In March 1945 the Ministry of Labour denied the NPL permission to advertise for staff and agreed only to the appointment of six senior staff and six assistants by personal recommendations. Consequently the appointments were not filled until Womersley returned from the United States in April 1945.

To coincide with Womersley's official appointment as Superintendent of the new Mathematics Division on 1 April 1945 the NPL issued press releases to *Nature* (Anon. 1945*b*) and the *Journal of Scientific Instruments* (Anon. 1945*a*). The press releases stated that a Mathematics Division was to be set up at Teddington and outlined Womersley's initial plans. From then on Womersley quickly began to assemble his staff and to equip the division. Although many of the early staff officially took up their appointments at the NPL in the autumn of 1945 and the spring of 1946 many had already begun work by the summer of 1945.[9]

Womersley, straying slightly from his original plans, initially organized the Division into *four* sections: Mathematical Statistics, Punched Card Machines, General Computing, and Analogue Machines. To form the first of these sections, Mathematical Statistics, Womersley invited several members of his staff from S.R. 17 to join him at the NPL some of whom came as early as June to begin work. The Punched Card Section also became operational during the summer of 1945. T. B. Boss was appointed to head the section after he had been transferred from the RAE (where he had spent some time working with R. A. Fairthorne and a Hollerith punched card machine installation[10]). Boss began to equip the section with punches, verifiers, a tabulator, and a sorter. By September, he had taken on four school leavers who immediately started punching cards for the Statistics Section. Before the end of the year these four girls had been given a short course in the use of Hollerith machines by BTM and subsequently trained new members of staff as they arrived.

Womersley had originally proposed that all the Admiralty Computing Service staff stationed at the NAO in Bath, where they had been evacuated to during the war, be transferred to Teddington to form the General Computing Section of the NPL Mathematics Division. However this proved to be impossible for two reasons. First, some of the staff wished to go elsewhere after their release from wartime duties in the civil service and secondly, the NPL was not prepared to take on the entire junior staff. Goodwin was offered the senior appointment in the General Computing Section and with him came L. Fox, P. H. Haines, F. W. J. Olver, H. H. Robertson, and J. Staton all from the ACS (Goodwin 1984). Goodwin arrived at the NPL in mid-August and was quickly followed by the others from Bath. Again several school leavers were taken on as junior computing staff. Womersley initially equipped the General Computing Section with several hand-operated desk machines, principally Brunsvigas, Marchants, and Fridens, and two electric machines, a Madas and a Monroe. Goodwin and the others from Bath brought with them unfinished ACS problems and work in the section began immediately.

During the autumn of 1945 Womersley decided that, as a stopgap measure before the construction of the large differential analyser could begin, the NPL Mathematics Division should take over responsibility for the Differential Analyser Group at Manchester University. After some negotiation the Differential Analyser Section of the NPL Mathematics Division was established on 1 January 1946 at Manchester University and not at Teddington. The Differential Analyser Section was the last of the four initial sections to become operational.

The most significant change Womersley made during 1945 to his original proposal resulted directly from his American tour earlier that year. Through his contacts with Hartree, Womersley had become aware of the importance of the ENIAC project[11] and had been given permission to visit the Moore School to observe the machine; Womersley was the first principle British visitor to the project (Goldstine 1972). Womersley was also aware of Alan Turing's work on computable numbers (Turing 1936) and, vaguely, of his work at the Government Code and Cypher School at Bletchley during the war.[12] Womersley contacted Turing offering him a post at the NPL to design and construct an electronic digital computer. Turing jumped at the chance. He took up his appointment in October 1945 and produced his now famous proposal for the development of an automatic computing engine (ACE) (Turing 1946).

Thus by the end of 1945 the NPL Mathematics Division had been set up and was 'functioning on a limited scale' with a staff of 22 (NPL 1942–7, 23 October 1945). The ACE computer project had also been initiated and was to dominate the work of the Division over the following decade.

ENDNOTES TO CHAPTER 7

1 The New York Mathematical Tables Project was set up in 1938 as part of the employment programme created by the US Government during the Depression. In 1942, after the United States had entered the war, the US Office of Scientific Research and Development (OSRD) set up the Applied Mathematics Panel to co-ordinate computation within government agencies. The Applied Mathematics

Panel took over responsibility for the Mathematical Tables Project and used the facilities to carry out computational work for a variety of government departments (Rees 1982).

2 Subjects later to become numerical analysis and computer science respectively.

3 Sir Charles Darwin (1887–1962) (FRS), grandson of Charles Darwin author of the *Origin of species*, took up the post of Director of the NPL in 1938 after a distinguished academic career as a mathematical physicist at Manchester, Cambridge, and Edinburgh.

4 Plummer Professor of Mathematical Physics at Cambridge.

5 See Chapter 5.

6 See Chapter 6.

7 During the war the SCS had made considerable use of the Customs and Excise Hollerith punched card installation at Buxton.

8 The inclusion of a Mallock Machine in this list probably owes more to Darwin as ex-chairman of the Cambridge Scientific Instrument Company, the manufacturers of this machine, rather than the success (or lack of it) which Maccoll at Cambridge or Duncan from the Ministry of Aircraft Production had had using the machine.

9 This discrepancy was due to the variation in release dates of personnel from their wartime postings.

10 See Chapter 9.

11 The ENIAC (Electronic Numerical Integrator and Computer) was developed at the Moore School of Electrical Engineering, University of Pennsylvania during 1942–5 by J. Mauchley and J. Presper Eckert. It was built to carry out ballistics calculations. However, from the ENIAC project developed the concept of an electronic calculating machine which stored instructions in the same way as it stored data, i.e. the stored program computer with which we are familiar today.

12 During the war Turing had been involved with the design and construction of code-breaking machines at the Government Code and Cypher School, Bletchley. See Hodges 1983.

8

The National Physical Laboratory Mathematics Division: a national computing centre

During 1946 the NPL Mathematics Division became fully operational. By the end of that year the Division had a staff of 50 divided into five sections: the General Computing Section, the Punched Card Section, the Differential Analyser Section, the Statistics Section, and the Automatic Computing Engine (ACE) Section. The Mathematics Division was operating more or less within the guidelines set out in the 1944 DSIR Interdepartmental Committee Report. From the description of a Central Mathematics Station as outlined in that report it might be assumed that the NPL Mathematics Division also embodied the concept of a computing centre defined by the characteristics given in Chapter 1. However, before such a conclusion can be reached it is important to examine the NPL Mathematics Division more closely.

THE NPL MATHEMATICS DIVISION 1946—1951

Computing machinery

With the exception of the Statistics Section the basis on which the Mathematics Division had been divided into sections had been the type of computing machinery used. The General Computing Section used desk calculating machines and a National 3000 Accounting Machine and was commonly called the Desk Machine Section. Part of the duties of the General Computing Section was to become familiar with, and hence advise on, all types of desk calculating machine. Initially, however, the number of desk machines at the NPL was small due to the difficulty of obtaining any equipment of this sort immediately after the war. Womersley had ensured that several machines were available by the time staff began to arrive at the NPL but T. Vickers, one of the early staff recalled that it was not easy to acquire many more machines:

> Originally machines were scarce or needed dollars (also scarce). We were helped by 12 Brunsvigas 'found' in Germany by our marauding forces. A 'survey' throughout the other 9 Divisions of the NPL found one or two more in 'cubby holes'. (Vickers 1984)

As Britain recovered from the wartime restrictions and shortages, the availability of calculating machines increased and the division was supplied with each new machine as it came on the market by His Majesty's Stationery Office (HMSO). The section tested each machine and reported on its usefulness; HMSO supplied calculating machines throughout the Civil Service.

The Punched Card Section hired their machines from the British Tabulating Machine Company Limited (BTM). Leasing, rather than purchasing punched card machines was normal BTM practice as the machines needed regular servicing and maintenance. At first the section was equipped only with the basic punched card machines – punches, verifiers, two sorters, two reproducers, an interpreter, and a tabulator – but gradually a substantial collection was built up. One of the section's early tabulators was the machine which G. B. Hey had had modified by BTM while working with Comrie on the 1936 investigation into the spacing of sugar beet.[1] This tabulator differed from the standard machine in that it was fitted with additional relays, extra distributors, and decimal rather than sterling counters. It was not, however, a reliable machine. It was old, much used, and, Hey (1983) suggests, once dropped off the back of a lorry. In 1948 an IBM Pierce Tabulator fitted with alphanumeric rather than numeric printing sectors from Bletchley Park was installed. Later, in 1951, an IBM 602A electromechanical calculator was installed which, along with the Bull 506 Multiplying Punch which arrived two years after, allowing the section to increase the amount of mathematical rather than data processing work it carried out.

Until November 1948 the NPL Mathematics Division Differential Analyser Section continued to use the differential analyser at Manchester University. In November 1948 the machine was moved to an old wind tunnel at the NPL and, as a result, required a great deal of attention to get it back into working order. (It has to be remembered that by 1948 the differential analyser was 10 years old and had been used for most of that time.)

The ACE section was different to the General Computing, Punched Card and Differential Analyser Sections in that it was initially set up to build, rather than simply exploit, a specific type of calculating machinery.

Personnel

Under Womersley the section heads were E. T. Goodwin—General Computing Section; T. B. Boss—Punched Card Section; J. G. L. Michel—Differential Analyser Section; E. C. Feiller—Statistics Section; and A. Turing—ACE Section.

The staffing of the General Computing Section (some of whom had transferred from the ACS with Goodwin) illustrated the type of work the section was asked to perform. First, it had a strong team of mathematicians, including Goodwin, Fox, and Wilkinson (part-time), concentrating mainly on numerical research. Secondly desk machine computations were carried out by a well-trained team of more junior staff led by T. Vickers. This part of the section had three main tasks; first to become expert in the choice of desk machines for a particular problem; second to exploit the National 3000 Accounting Machine to its full potential for mathematical problems; and third to undertake computational work submitted to the section. Where jobs required non-standard techniques or research the mathematicians of the section supplied the necessary expertise while the desk machine operators carried out the bulk of the work.

Other than T. B. Boss, and his deputy Florence Rigg, the staff of the Punched Card Section consisted primarily of school leavers trained specifically to use the NPL punched card machines for scientific work. They worked in teams of one to three depending on the complexity and size of each job. Each team took jobs from their planning, plugging, and running stages through to the presentation of results. The mathematicians from the General Computing Section also contributed to the work of the Punched Card Section, in particular devising numerical methods applicable specifically for use with punched card machines (see, for example, Fox *et al.* 1948).

In contrast to the well-trained, close-knit teams of the General Computing and Punched Card Sections, staffing in the Differential Analyser and Statistics Sections was more problematic. When the NPL took over responsibility for the Manchester differential analyser in 1946 the existing differential analyser staff, originally under Hartree, were officially transferred to the NPL. However, as in the case of the relocation of ACS staff, after their release from wartime posting some moved elsewhere. Others who were later recruited to the Differential Analyser Section moved to other Mathematics Division sections as their interests changed to numerical analysis or the ACE project. By 1949 the Differential Analyser Section had a staff of only four which was not sufficient to keep the machine in continuous operation.

The staff of the Statistics Section had originally transferred from the Ministry of Supply as an autonomous group. The work they carried out was also very different to that performed by the other sections. These two factors resulted in the Statistics Section being perceived by other Mathematics Division staff as being separate from the rest of the division. This attitude was further aggravated by the fact that the Statistics Section had a very high turnover of staff which meant that professional, and personal, relationships were not built up.

Initially the staff of the ACE Section consisted only of Turing and J. H. Wilkinson whose work was split between the ACE and the General Computing Sections.[2] Wilkinson had worked in the Armament Research Department during the war. At the end of the war Wilkinson was not released from service immediately and was assigned to the NPL Mathematics Division. His intention was to return to Cambridge to undertake mathematical research. However the experience of working with Turing and Leslie Fox from the General Computing Section 'fired [him] with so much enthusiasm for the computer project and so heighted [his] interest in numerical analysis that [he] gradually abandoned the idea' (Wilkinson 1971, p.139). In September 1946 Mike Woodger[3] joined the ACE Section to be followed in January 1947 by H. D. Huskey, an American, who had been involved in both the ENIAC and the EDVAC projects, and who was to spend a year at Teddington. Although the staff of the ACE Section was small it was well able and qualified to undertake the task of designing the ACE.

Provision of a computing service

The main motivation behind setting up the NPL Mathematics Division had been to provide a computing service to government departments, industry, and the universities. There is no doubt that the General Computing and Punched Card Sections fulfilled that role. These sections were well equipped in terms of both staff and machinery to carry out this role. The range of users was varied and problems brought to the sections ranged from the small to the large, and from the routine to the complex. The Differential Analyser Section undertook some service work but this was very limited due to the staffing problems of the section. However the section did undertake a reasonable amount of advisory work for existing Manchester users and institutions who wished to build or install a differential analyser in their own premises. The section also gave advice on other types of analogue computation including nomograms and harmonic analysers.

The situation regarding the Statistics Section was much the same as in the Differential Analyser Section. The Statistics Section used the punched card and desk machines from the other Mathematics Division sections to carry out the small amount of service work it undertook. However, much of the section's work was advisory and came from a variety of sources requiring anything from 'half an hour's oral service to investigations requiring extensions of existing theory and detailed analyses covering several weeks' work' (NPL 1949*a*).

The service and advisory work of the NPL Mathematics Division taken as a whole flourished during the 1946–51 period and entirely justified the creation of a National Computing Centre. A list of the division's users during that time is given in Appendix 4A.

Computing machinery research

Acting as a computing service was one of the principal roles of the NPL Mathematics Division and a source of income. The other distinct role played by the Mathematics Division was that of a centre for research into computing machinery. In March 1946 the NPL Executive Committee drew up a list of computing machinery development projects which the Mathematics Division was to undertake. The Committee listed the projects in the following order of priority:

(1) ACE machine;
(2) a small Fourier synthesis machine for use by X-ray crystallographers in their own laboratory;
(3) a production model differential analyser;
(4) an O/C Osser-Jofeh[4] machine to combine the functions of
    (a) Fourier analysis,
    (b) Fourier synthesis, and
    (c) an electric isograph;
(5) a large differential analyser.

From the list only the ACE and large differential analyser projects were undertaken and of these the ACE was to prove by far the most significant.

A thorough description of the design and construction of the Pilot ACE and the ACE is beyond the scope of this discussion where only the circumstances surrounding the machines will be considered. Descriptions of different technical aspects of the ACE project have been given elsewhere (Campbell-Kelly 1981; Huskey 1984; Lavington 1980; Wilkinson 1975; Woodger 1958).

In February 1946 Turing's proposal for an Automatic Computing Engine (Turing 1946) was presented to the NPL Executive Committee accompanied by a supporting document written by Womersley (NPL 1942–7, 19 March 1946). The papers were discussed by both the Executive Committee and the DSIR Advisory Council before Treasury approval for the project was granted in the summer of 1946. During 1946 Turing, Wilkinson, and Woodger worked on the logical design of the ACE. However as Turing was the only member of the ACE section to have any experience with electronics it was expected that the actual construction of the machine would be done outside the NPL (Huskey 1984 and Woodger 1977). Consequently arrangements were made for the Post Office Research Establishment at Dollis Hill to develop mercury delay lines to be used as memory units for the ACE. Because of research commitments of its own the Post Office was reluctant to undertake any further development work for the NPL. Darwin, on behalf of the Mathematics Division, then approached F. C. Williams from the Telecommunications Research Establishment and later Manchester University, and M. V. Wilkes from the Cambridge Mathematical Laboratory both of whom were involved in separate computer projects at that time. Although Womersley conducted negotiations with both Wilkes and Williams no contract for the construction of the ACE was signed. Williams wished to pursue his own research separate from Turing and Turing's design differed considerable from the machine Wilkes was intending to build. Therefore by April 1947 it had been decided that the NPL should construct its own machine (Woodger 1977).

After this decision had been taken Huskey proposed that rather than build a full-size machine the Mathematics Division should construct a test assembly to determine the feasibility of the design. However, because of the limited experience in the use of electronics within the Mathematics Division, Darwin took the responsibility for all hardware construction out of their hands (Huskey 1984). The original idea of constructing the test assembly was subsequently abandoned in favour of a plan to build a 'small scale pilot model' of the ACE in the newly formed Electronics Section of the NPL Radio Division (NPL 1949*b*). The Electronics Section later became independent of the Radio Division in 1948.

Turing, never in favour of building a pilot model preferring instead the immediate construction of a full-size machine, left the NPL in October 1947 for a sabbatical year at Kings College, Cambridge. He never returned to the NPL but took up a post at Manchester University to work on the computer project there.[5] Wilkinson took over from Turing as section head and began to recruit more staff to the project.

Although the Mathematics Division's main responsibility was for logical design and programming of the ACE, in August 1948 four members of the ACE Section (Wilkinson, Woodger, Alway, and Davies) were seconded to the Electronics Section to work on the construction of the pilot machine. Throughout 1949, 1950, and 1951 work continued both on the construction of the Pilot ACE in the Electronics Section and on programming in the Mathematics Division. Although the machine carried out its first automatic calculation in May 1950, it was not completed until late 1951 after which it was moved to the Mathematics Division in early 1952. While waiting for delivery of the machine the General Computing Section had prepared for its arrival by producing a number of sub-routines for performing factorization, quadrature, integration of simple differential equations, and the solution of simultaneous equations. Thus when the Pilot ACE was finally installed it was immediately put into service (Plate 10).

Unlike the ACE Section the Differential Analyser Section did not design and assist in the construction of their own machine, although up until late 1948 Michel, head of the section, had been working on the design of a 24-integrator differential analyser to be built at the NPL. In March 1949 it was decided to install a 20-integrator machine being built in Germany by Schoppe and Faeser. Consequently the Differential Analyser Section staff spent much time in Germany during 1949–54 working on the machine. Once the differential analyser units had been delivered to the NPL, staff became very involved in the wiring together of the separate units, despite Metropolitan Vickers being under contract to carry out this work. Thus, while there was a great deal of work preparing for the installation of the 20-integrator machine and its wiring, no fundamental hardware research was carried out by the section staff.

Numerical research

Research into what would now be called numerical analysis also flourished in the NPL Mathematics Division. The mathematicians of the General Computing Section, namely Goodwin, Fox, and Olver, carried out research which reflected their own interests, the needs of the Royal Society Mathematical Tables Committee, and, later, techniques applicable to the Pilot ACE. Because of the distinction between mathematicians and desk machine operators the senior staff were not overwhelmed by service work and could therefore devote much of their time to research. Areas in which work was carried out included the solution of simultaneous equations, relaxation techniques, integral equations, algebraic equations of high orders, polynomial approximations, and table-making.

A thorough discussion of the numerical analysis research carried out by the General Computing Section would be out of context here. However, much of the numerical work of the General Computing Section was published in journals such as *Mathematical tables and other aids to computation*, *Quarterly journal of mechanics and applied mathematics*, *Proceedings of the Royal Society*, *Philosophical transactions of the Royal Society*, *Proceedings of the Cambridge Philosophical Society*, and others. By publishing their work and contributing to the rapidly expanding field of numerical analysis the individual mathematicians of the NPL became well known in their own right and gave rise to the NPL's international reputation as a leading centre for numerical analysis.

Library facilities

As part of their role as 'experts' in the computational field the NPL Mathematics Division also maintained a specialist library. Not only did the library hold published tables and literature but internal reports and papers were collected. As a government establishment the NPL Mathematics Division was also in a position to hold unpublished papers originating within other government departments which had computational interests. For example, papers concerning punched card techniques used at the RAE were held by the division's library.

The NPL Mathematics Division was a well-used computing service bureau and advisory service, it was heavily involved in computer research and research in numerical analysis, and the staff published the results of their work in well-known scientific journals and by holding a computer conference in 1953 (NPL 1954*a*). The division also maintained a specialist library. In other words the NPL Mathematics Division was a national computing centre as described by the criteria outlined in Chapter 1. In 1952 the Pilot ACE became operational and, although the NPL Mathematics Division continued to function as a national computing centre for some years, the emphasis of the work of the division changed.

REORGANIZATION OF THE MATHEMATICS DIVISION: CAUSES AND RESULTS

In September 1950 Womersley resigned as Superintendent of the NPL Mathematics Division in order to take up a position with BTM the punched card machine manufacturer. In February 1951 E. T. Goodwin succeeded Womersley as Superintendent. Goodwin had to immediately apply himself to the imminent installation of the Pilot ACE. It was obvious that more staff would be needed in the ACE section to keep the Pilot ACE fully utilized. Because of the government freeze on recruitment to the Civil Service the additional staff necessary had to be found from within the division itself. Goodwin felt that the Statistics Section 'had never been fully integrated into the rest of the Division, (Goodwin 1984). This was, to a large extent, true given the type of work the Statistics Section performed and the high turnover of its staff. In the original 1944 DSIR Interdepartmental Technical Report the distinct nature of the statistical work of the division had been acknowledged:

> if the expansion in statistical work that is adumbrated should take place, it might be advisable later to split off the Statistical Section from the Mathematical Division. (DSIR 1944, p.6)

For these reasons Goodwin came to the conclusion that 'the only way we could recruit to man our electronic computer was to find a niche in another Ministry for our Statistics Section while retaining their posts' (Goodwin 1984, pp.6–7). As a result, the Ministry of Supply announced in 1951 that it was going to re-establish its own statistical group and stated that it would accept the transfer of the NPL Statistics Section to this group. The Ministry of Supply was expected to continue the service and consultancy role played by the NPL Statistics Section and thereby fulfil the statistical needs of other ministries. The Mathematics Division therefore gained approximately six posts to which it recruited personnel for the ACE Section.

In 1952 the Pilot ACE went into regular service and Goodwin restructured the division in February of that year. Goodwin divided the staff of the Mathematics Division into three groups: High Speed Computing supervised by Fox, Low Speed Computing supervised by Boss, and Mathematics supervised by Wilkinson.

The first of these groups, Fox's High Speed Computing Group, was divided into the ACE and Differential Analyser Sections. The ACE Section now consisted primarily of hardware development and operational staff (D. W. Davies, York (part-time), and Vickers) and junior computing staff. Vickers was responsible for the staffing and day-to-day running of the ACE. The Differential Analyser Section remained intact. When the 20-integrator differential analyser came into operation in 1954 Vickers took charge of the programming staff of both machines. This section prepared and ran programs for the ACE and differential analyser as part of the growing NPL computing service.

The Low Speed Computing Group was made up of the Desk Machine and Punched Card Sections. It was headed by Boss who also took responsibility for the computing machinery advisory service the division provided to the Treasury and HMSO. The Desk Machine and Punched Card Sections continued to work on computing service jobs which were unsuitable for transfer to the ACE. The Desk Machine Section carried out small calculations which would have taken a disproportionate amount of time to program for the value of the results or very complex calculations which involved a very large amount of complex programming. Other jobs, such as those involving repeated calculations on a very large amount of data or those which required much data manipulation, were more easily performed using punched card machines than on the Pilot ACE which had, initially, only 320 32-bit words of memory. Some numerical research was carried out in the Desk Machine Section by the mathematicians who remained.

The Mathematics Section constituted the senior programming staff of the former ACE section and the remainder of the division's numerical analysts. It formed a 'pool' of scientific, rather than assistant, staff on which the High Speed and Low Speed Computing Groups could draw. It was headed by Wilkinson and included York (part-time), H. H. Robertson, M. Woodger, J. G. Hayes, G. G. Alway, and B. Curtis. The work of the section over the following two to three years concentrated primarily on building up a library of subroutines for the Pilot ACE rather than on other types of numerical work. The Reports of the NPL covering the period 1952–5 indicate that Goodwin was concerned that the Mathematics and ACE Sections were spending too much time on service problems and not enough on basic research. In 1954 Goodwin tried to solve this problem by formally dividing the ACE section and that part of the Mathematics Section most concerned with electronic computers into two separate groups; a research group and an operational group. This allowed the more routine problems to be passed directly to the latter group which had now gained programming experience and hence left the senior staff more time to devote to research.

From the original memo outlining this organizational change (Goodwin 1952) it is clear that Goodwin intended the senior staff of the Desk Machine Section to eventually transfer to the Mathematics Section and the junior staff to move to the ACE and differential analyser programming teams as electronic computers became more dominant.

AFTER THE PILOT ACE: 1952–1957

The installation of the Pilot ACE initiated a substantial increase in the amount of work which came into the NPL Mathematics Division and it became necessary to introduce a shift system in order to fulfil the demands on machine time. Appendix 4B lists some of the more important Mathematics Division users during the 1952 to 1957 period. Several customers made regular use of the NPL facilities particularly the RAE, English Electric, the Ordnance Survey Department, the Ministry of Supply, and, later, the United Kingdom Atomic Energy Authority (UKAEA).

From 1948 English Electric had been working closely with the NPL Electronics Section in order to produce an engineered version of the Pilot ACE, the DEUCE. The DEUCE, was delivered to the NPL in 1955 and prompted a further increase in custom for the High Speed Computing Group. At the same time the NPL lost two of its heaviest customers when a DEUCE was delivered to the RAE, which meant that Farnborough staff no longer used the NPL facilities, and English Electric kept two DEUCE machines for its own use. Despite the loss of these two regular users there was a steady increase in the number of users between 1955 and 1957. The DEUCE was also put into operation in the evenings and on weekends in order to cope with the demand of its services.

As demand for electronic computing grew the work load of the other NPL service sections began to decline. The 20-integrator differential analyser, although considerably more powerful than the Manchester machine, was never used as heavily as the Pilot ACE or the DEUCE. By 1953 it had become obvious to Goodwin that the future lay in the development of electronic computers and methods for use with them, and not in the construction of analogue machines. Goodwin writes,

> I think that it is also fair to say that I was much less enthusiastic about the Differential Analyser than Womersley was. I was quite convinced that the Electronic Computer would soon oust the analogue machines and, though Jack Michel worked manfully to keep the D.A. fully operational and doing useful jobs the writing really was on the wall soon after I took over. (Goodwin 1984, p.7)

Use of the differential analyser stopped in 1958.

After the installation of the Pilot ACE and the DEUCE the number of Desk Machine Section users was drastically reduced due to the calculations traditionally performed by the section being undertaken by the High Speed Computing Group. The number of staff in the section also diminished as they gradually transferred to Vickers' programming team. *Ad hoc* or exploratory calculations for other Mathematics Division sections or other NPL Divisions soon became the mainstay of the Section's work. In contrast to this the work of the Punched Card Section, surprisingly, increased.

Although the High Speed Computing Group took over much of the work formerly carried out by the Punched Card Section, particularly the matrix manipulation work, during 1953–6 the section performed an increasing amount of service work. This enormous influx of work came from a single user—the United Kingdom Atomic Energy Authority (UKAEA). As part of their research programme the UKAEA used the Monte Carlo technique to simulate the behaviour of neutrons under certain conditions. Although the Atomic Energy Research Establishment (AERE) at Harwell acquired a punched card installation in 1954 equipped with IBM 602A and 626 calculators and operated by staff trained at the NPL, there was still a large demand for this type of work from the UKAEA. The installation in 1955 of an IBM 626 calculator at the NPL was directly attributable to this work and was followed a year later by a second machine paid for by the UKAEA. In 1957 the Atomic Weapons Research Establishment (AWRE) at Aldermaston installed an IBM 704 computer and a year later a Ferranti Mercury arrived at Harwell. These machines took over the Monte Carlo and other calculations the NPL had been performing on behalf of the UKAEA and the work load of the Punched Card Section fell dramatically at this point.

In 1954 the NPL Electronics Section and the Control Mechanisms Section of the Metrology Division were combined to form the Control Mechanisms and Electronics Division (CME). The role of the CME Division was to develop both computational and automatic control mechanisms. Both parts of the new Division had been involved in the development of computing hardware for the Mathematics Division; the Electronics Section worked on the Pilot ACE and the DEUCE, and the Control Mechanisms Section had been heavily involved with the installation and testing of the German differential analyser and the design of an automatic curve follower for the machine. In 1954 the CME Division began work on the full-size ACE. This effectively took almost all research on computing machinery out of the hands of the Mathematics Division. Consequently the type of research carried out by the Mathematics Division concentrated on numerical analysis, applied mathematics, and programming.

As the Mathematics Division became a recognized computer centre rather than a computing centre the advisory role it played took on a new dimension when its opinion began to be sought by universities considering setting up their own computing laboratories. It also became involved, as part of the NPL's work on standards, in devising acceptance tests for computers delivered elsewhere. The first of these acceptance tests was for a Ferranti computer being delivered to the Royal Dutch Shell company (NPL 1955).

In 1952 the Mathematics Division had been reorganized into the High Speed Computing Group, the Low Speed Computing Group, and the Mathematics Section. In 1957 the organizational structure was changed for the second time. The new sections were the Numerical Methods Group, the Applied Mathematics Group, and the General Computing Group.

The Numerical Methods Group, under F. W. J. Olver, consisted primarily of the old Mathematics Section and those mathematicians previously based in the High Speed Computing Group. Its main areas of interest were to be numerical analysis, including error analysis, the development of numerical techniques suitable for application to electronic computers, and the continuation of the work on both the Royal Society and NPL Mathematical Tables Series. This group, with staff from the CME Division, was responsible for software research on the DEUCE after its installation in 1955. For example, an important programming scheme, Generalized Interactive Program (GIP), was developed for linear algebra problems; GIP became one of the most common DEUCE programming systems in the late 1950s (Campbell-Kelly 1980*c*).

The second group, Applied Mathematics headed by Michel, was created from the Differential Analyser Section. It was not, however, responsible for the operation of the differential analyser. Rather, most of their work dealt with the solution of differential equations and other forms of applied mathematics and theoretical physics, for example, developing methods for the solution of integro-differential equations involved in problems of nuclear scattering.

The DEUCE, Desk Machine, Punched Card Sections, and the operational side of the Differential Analyser Section were brought together under Vickers to form the General Computing Group. All the service sections were therefore collected into one group and not separated into distinct sections by machine type. This arrangement reflected the increasing dominance of electronic computers and the subsequent decline of other computing machines.

The reorganization, coupled with the earlier loss of responsibility for hardware research to the CME Division, changed the emphasis of the Division's work. It no longer fulfilled the recommendations of the 1944 Interdepartmental Committee's Report to carry out research into new computing machinery. But the environment in which the Division operated had also changed. It was no longer necessary for a national computing centre to develop electronic computers for they were being manufactured and sold commercially by English Electric, Ferranti, Lyons, and IBM.

As other research establishments and universities began to install computers the amount of service work required by these large NPL users fell while the demand and need for numerical research continued. Although the General Computing Group continued to receive incoming work for several years it gradually became a local computing service for other NPL divisions and not a national facility. The main work of the division was now concerned with numerical and programming research. Its interests were turning it into a national mathematical centre rather than a national computing centre.

ENDNOTES TO CHAPTER 8

1 See Chapter 4.

2 Officially Wilkinson spent half his time in the General Computing Section and half with Turing in the ACE Section. In practice, however, Wilkinson spent most of his time working with Turing, only exercising his right to go elsewhere when Turing's mood made it advisable to stay out of the way (Wilkinson 1971).

3 Woodger has since become well known for his work on Algol.

4 Details of an Osser-Jofeh machine are not available in the usual sources concerning calculating machines of the late 1930s and early 1940s. It is assumed that someone on the NPL Executive Committee, possibly Darwin or Hartree, had some connection with the machine hence its inclusion on this list.

5 See Chapter 10.

9

Post-war computing service centres

The NPL Mathematics Division was set up to act as a national computing centre to serve most of Britain's computing needs. It epitomized the centralization concept both by providing computing services of all kinds and by carrying out research into new computing machinery and numerical techniques. The existence of the NPL Mathematics Division did not prevent several other institutions from developing computing centres of their own. Perhaps the most well-known British scientific computing service centre, other than the NPL Mathematics Division, was the Scientific Computing Service run by L. J. Comrie.

THE SCS 1945–1950

Before and during the Second World War the SCS had been very much in demand by the armed services and other government departments. The end of hostilities and the demobilization of scientific staff meant that demand for computation fell dramatically in the immediate post-war period. The SCS could no longer depend on receiving urgent requests for tables or calculations from the Admiralty, Air Force, Army, or Ministry of Supply. The establishment of the NPL Mathematics Division and the computing service which it offered also reduced the amount of work the SCS was asked to perform. The SCS had lost its monopoly in the computing service field. Consequently the type of work which the SCS undertook began to change from a wide range of jobs of varying complexity to work involving, predominantly, table-making and large analyses for X-ray crystallographers.

By far the largest job undertaken by the SCS during the post-war period was the revision of *Chambers's mathematical tables* (Comrie 1948*a*, 1949, 1950). Towards the end of the war W. and R. Chambers Ltd. approached Comrie on the recommendation of A. C. Aitken, the mathematician, and W. H. M. Greaves the Scottish Astronomer Royal (and also Comrie's brother-in-law). Chambers asked Comrie to report on their 7-figure tables and to consider the possibility of updating them. After reviewing the tables Comrie declared that the tables and existing printers' plates were not worth preserving and undertook a complete revision of the tables.

Over the following three years much of the work performed by the SCS involved the computation and proof-reading of the revised *Chambers's mathematical tables*. Apart from typographical alterations and the recomputation of the tables using National Accounting machines, Comrie introduced several major changes to the content of the tables. He argued that 7 figures were rarely used by general computers because, in the majority of cases, 5 or 6 figures would suffice. Consequently Comrie reduced the table to 6 figures. He also omitted many of the specialist nautical, astronomical, and survey tables which were available elsewhere and included tables of the exponential, hyperbolic, inverse hyperbolic, and inverse circular functions which had come into general use since *Chambers's mathematical tables* was first published in 1884. In all Comrie and his staff computed and proof-read five volumes of tables: two volumes constituted the main tables, one containing the natural values of functions the other the logarithmic; one volume titled *Chambers's shorter six-figure mathematical tables*; and two volumes of 4-figure tables.

Comrie was very concerned about the accuracy of the tables and encouraged his staff to detect errors by offering monetary rewards. For example, he paid 2*d*. for every error found in the first proofs and charged 6*d*. for every error missed. For errors found in plate proofs he offered 5*s*. By the time the tables were published in 1947 Comrie was so confident of the absence of errors exceeding ± 0.51 in the last decimal that he offered a £5 reward for their detection (Comrie 1948*a*). Comrie himself knew of less than five errors in the tables and these he left uncorrected as 'an uncomfortable trap for any would be plagiarist' (Comrie 1950, p.vi).

The other major computation carried out by the SCS during the late 1940s was a crystallographic Fourier analysis for Bragg and Perutzi at Cambridge University on the molecular structure of haemoglobin. During the war the SCS had used IBM tabulators at the Milk Marketing Board at Cirencester to perform the Fourier analysis required for X-ray crystallography. The use of American machines was critical in this work because they allowed direct subtraction from the card. British tabulators, supplied by BTM, did not. Through BTM, the SCS acquired an IBM tabulator which Comrie installed in a converted wool shop in Wimbledon. By using this machine and other punched card machines in different installations all over London, the SCS completed this and several other crystallography jobs (Gittus 1983).

Because information regarding the post-war activities of the SCS is scarce (the principle sources are the personal recollections of Comrie's staff) it is difficult to list the other computations carried out by the company during this period with any degree of accuracy. What is clear, however, is that apart from the recomputation of *Chamber's mathematical tables* and the few large crystallography jobs which were undertaken, the SCS performed only small calculations. Because of the decline of the computing service side of the business, the publishing work of the SCS began to take on a greater importance.

The most important book published by the SCS was Fletcher, Miller, and

Rosenhead's *An index of mathematical tables* which appeared in 1946. During the war Comrie had become aware of a project being undertaken at Liverpool University by A. Fletcher, J. C. P. Miller (who later came to work for Comrie), and L. Rosenhead to compile a practical guide to published mathematical tables. The guide included information on the numbers of figures given in each table listed, the size of the interval, whether or not differences were provided to facilitate interpolation, authorship, and overall quality of the tables. The tables were grouped together in order of the function tabulated and also listed by author in a comprehensive bibliography. Comrie himself had been working on a similar project for a number of years but had not completed the work. Consequently Comrie offered to publish the *Index*. After many delays, due to the amount of war-related work the printers were committed to, the first edition of the *Index of mathematical tables* appeared in 1946 and received enthusiastic reviews throughout Britain and the United States. Almost immediately an expanded second edition was begun which included an important section listing the known errors in published tables based on Comrie's own work in this area. The second edition was published in 1962 and Comrie's name apeared posthumously as a co-editor.

In addition to acting as a publisher for the *Index of mathematical tables* and other tables, the SCS also operated as a bookseller specializing in mathematical tables and relevant literature. The SCS held all the tables published by the British Association Mathematical Tables Committee from 1931 onwards, the Annals of the Computation Laboratory at Harvard Unviversity, the tables published by the United States National Bureau of Standards Mathematical Tables Project, and over 70 other related publications.

In addition to his work with the SCS Comrie continued to write and lecture on computing methods and machines and hence built up the image of the SCS as a computing centre. In October 1945 he visited the United States to chair a conference organized by the National Research Council Committee on Mathematical Tables and other Aids to Computation at the Massachusetts Institute of Technology. At the conference Comrie delivered a paper on 'The scientific application of commercial calculating machines' which gave a historical account of the development of calculating machines and their use for scientific work (Comrie 1946*b*). While in the United States, Comrie visited the Watson Computing Laboratory at Columbia University, the US Naval Observatory, the US Coast and Geodetic Survey Office, the National Bureau of Standards, Harvard University, and the Moore School of Electrical Engineering.

Although he visited both the Automatic Sequenced Controlled Calculator at Harvard and the ENIAC at the Moore School of Electric Engineering in Pennsylvania, Comrie remained cautious about these new developments. Comrie was well known for making disparaging remarks about those who constructed large, expensive, specialized machines. In his view most computing work could be more economically carried out using desk calculators and commercial adding and listing machines. For example, following a paper read by Beevers before the Physical Society in 1939 which described a machine to perform the Fourier series summations required in X-ray crystallography analyses, Comrie made the following remark which summarized his views on specialized computing devices and the benefits of centralized computation. He said

> I have always advocated, as a policy, that the capabilities of existing commercial machines should be fully exploited before the laborious and costly task of designing and making special apparatus is embarked upon. In this case the nature of the problem points to the use of punched-card machines, which have been successfully applied to Fourier synthesis in the summation of harmonic terms in the moon's motion. I am not convinced of the curt dismissal of these machines. (Comrie 1939*a*, p.664).

Likewise in 1944 he dismissed the Mallock and Wilber simultaneous equation solvers by stating that the designers were 'suffering from the common delusion that such solution is difficult' (Comrie 1944, p.134). Two years later he reiterated this point of view in a statement made during the discussion of a paper read to the Royal Statistical Society by H. O. Hartley, a senior employee of the SCS, describing methods of mechanical computation. Comrie is reported to have said:

> He was glad that no one had got up that evening to talk about making special machines, because, before they made special machines to do things, it was well to explore the possibility of existing machines. (Hartley 1946, p.177)

Comrie was much less critical of the large calculators and computers which began to appear the end of the war, possibly because they were not specialist machines capable of only a few limited applications but were more general purpose devices. Comrie did, however, gently question whether the expense of the Automatic Sequence Controlled Calculator built by Howard Aiken at Harvard was justified when the main published examples of the machine's work could have been performed using desk calculators (Comrie 1946*a*, p.568).

Through his acquaintance with Womersley, Comrie was aware of the work of the NPL Mathematics Division and the circumstances in which it was set up. Indeed there is evidence to suggest that Comrie and Womersley met on several occasions to discuss what role the NPL Mathematics Division should play in the scientific community (Goode 1983; Miller 1989). Comrie had some very definite ideas about how he felt the NPL Mathematics Division should operate. In his view the NPL Mathematics Division should have been the national centre for the development of computing methods and machines. Comrie envisaged the SCS complementing the work of the NPL by providing a computing service for the NPL and the wider scientific community in government departments, the universities and industry (Woodger 1977). However the flow of work from the NPL to the SCS which Comrie expected never really began; instead the NPL Mathematics Division developed its own computing service as the 1944 DSIR Committee had envisaged.

Despite the failure of Comrie's plans for the SCS to complement and not duplicate the work of the NPL Womersley did his best to assist Comrie when the SCS began to get into financial difficulties because of a lack of incoming work. For example the NPL Mathematics Division contracted out one piece of work on the SCS in 1946. Comrie was asked to undertake the computation of tables of Weber parabolic cylinder functions for a fee of £1200 (DSIR 1946). Again in 1949 Womersley tried to put the SCS on a more secure footing by suggesting to Darwin, Director of the NPL, that 'in view of possible future emergencies it is in the National interest that Scientific Computing Service Ltd. should be kept going' (Womersley 1949*b*). Womersley implied that government computing resources had not yet reached a level where they could independently cope with the demand created by a future war or national emergency. Womersley proposed that the NPL support the SCS by either subcontracting to the SCS or turning away suitable work and suggesting that the client use the SCS. Womersley argued that this scheme had the added advantage that the NPL's own computing staff could then be released from routine work in order to work on programming the Pilot ACE when it was finally installed. Apart from the ethical questions relating to whether or not a commercial concern should be used to perform work for DSIR research establishments Womersley did not have a chance to implement his plans as he left the NPL before the Pilot ACE was installed. However the example does serve to illustrate the close connections which existed between Comrie and Womersley.

From 1948 onwards the success of the SCS declined rapidly. The principle reason behind this decline was the relatively low demand for computation (caused partly by the existence of the NPL Mathematics Division) but the situation was aggravated by Comrie's deteriorating health. Because Comrie's health was failing others had to take over the administration of the company but it was difficult for any one person to step into Comrie's shoes as an expert in computation and typography and as an astute businessman with a good reputation and a large circle of acquaintances. Comrie died in December 1950 9 months after being elected a Fellow of the Royal Society. The SCS never regained its reputation as an important computing centre.

THE ROYAL AIRCRAFT ESTABLISHMENT, FARNBOROUGH

The only government establishment, apart from the NPL, to develop an important computing centre during the late 1940s was the Royal Aircraft Establishment (RAE) at Farnborough.

A. G. Pugsley, head of the Structural and Mechanical Engineering Department of the RAE, had been trying to improve computing methods since the mid-1930s,[1] but it was not until 1944 that the Ministry of Aircraft Production granted the necessary permission and finance to install Hollerith punched card machines at Farnborough. The machines were installed in August 1944[2] and became the basis of a Computing Laboratory to serve the Structures Department of the RAE. The Computing Laboratory was initially equipped with several desk calculators, a National Accounting Machine Class 3000, and the punched card installation which consisted of a sorter, a collator, a reproducer, an IBM 601 multiplying punch, and a E6/6 tabulator. The E6/6 tabulator was an old BTM model which did not have the facility to transfer numbers between registers. This machine was soon replaced with a Senior Rolling Total Tabulator which had inter-register transfer as a standard feature and allowed the RAE to carry out more complex work such as matrix multiplication.

R. A. Fairthorne was responsible for running the Computing Laboratory which was housed in an army hut at Farnborough, and at first T. B. Boss assisted Fairthorne in setting up the new installation. Boss had already been appointed head of the Punched Card Section of the NPL Mathematics Division and his work at the RAE gave him valuable experience in the use of punched card machines for scientific work. In September 1945 Boss left the RAE to take up his position at NPL and Fairthorne acquired several more members of staff to run the laboratory. In 1948 a new Mathematical Services Department was set up at the RAE under S. H. Hollingdale. The new department was made up from the separate computing facilities in different RAE departments principally Fairthorne's Computing Laboratory and several analogue instruments which had been in use in the Aerodynamics Department.

The Mathematical Services Department was set up to provide a centralized computing facility for the RAE as a whole. It maintained a good variety of calculating equipment and had a staff trained to carry out computations presented to it. Despite the fact that the Mathematical Services Department was set up specifically to act as a computing centre for the RAE it did not successfully achieve this objective. Without the co-operation of the research staff of the RAE centralization of computational work was impossible. Fairthorne observed that

> Departments, even without the security restrictions within a service department, are ever watchful of extra-departmental infiltration. This made centralized computing rather ineffective. . . . You can't do applied maths or applied computation properly without knowing quite a lot about the application. At any time this is a problem. In an institution that, rightly, had very tight security on much of its projects, and was fragmented by departmental chauvinism I had to exercise much tact and know a lot more than I was supposed to in order to get any work done at all. (Fairthorne 1983*b*, pp.1–2)

To try to dispel the reluctance on the part of RAE staff to allow others to perform their numerical work and encourage them to make use of the considerable computing power available to them, Fairthorne and his staff prepared a report which was intended to explain how the RAE punched card installation could be applied to a number of problems (Staff of the Mathematical Services Department 1952).

The report outlined the basic principles of punched card machines and gave an indication of how they could be applied to scientific work. The report then gave several examples of both successful and unsuccessful calculations which had been performed using the RAE punched card installation and tried to analyse why some jobs had been less successful than others. The report also put forward ways in which the methods used could have been improved if Mathematical Services Department staff had been given some background of the physics behind the problem or had been consulted on what format experimental readings should have been taken. One of the most successful jobs which the RAE punched card installation performed, an analysis of flight recorder records, owed much of its success to the type of interdepartmental co-operation being advocated. Because Fairthorne was personally acquainted with the person in charge of the investigation, he was able to arrange for data to be transcribed directly on to punched cards and hence not only eliminate one stage of the process but also reduce the danger of transcription errors entering the computation. In the published form of the report (Fairthorne *et al.* 1958) the value of consulting the Mathematical Services Department was again emphasized by the example of the flight recorder analysis job. The report noted that:

> For this job all people concerned co-operated before actual experiments were done. Thus the recording instruments themselves, the records and their monitoring and measurement, the analysis and its final presentation were all designed to match. The total time from trial to publication was reduced to less than one-thousandth of that previously needed. (Fairthorne *et al.* 1958, p.24)

In addition to being internally distributed, the report was published as part of the Aeronautical Research Council's Reports and Memoranda series with the intention of bringing the RAE punched card installation to the attention of all those engaged in aeronautical research. Because of administrative problems, however, the report was not published until 1958 (Fairthorne *et al.* 1958) by which time it was largely obsolete as the RAE had by then installed a DEUCE computer.

Because the RAE punched card installation was not used to any great extent by RAE staff, the Computing Laboratory, and later the Mathematical Services Department, took on outside work. Some of this work, including jobs for the Building Research Station and the Cambridge Language Unit, came to Fairthorne through BTM which passed on the more mathematical problems received by its service bureaux. The RAE punched card installation was also used by the Royal Society Mathematical Tables Committee (RMSTC) in 1949 to prepare tables of binomial coefficients for volume 3 of the Royal Society Mathematical Tables Series (Miller 1954; Woollett and Simm 1949). The National Accounting Machine was also used for outside table-making work. The most well-known example of this was the calculation of eigenvalues for a portion of the Hilbert Matrix for J. C. P. Miller (Fairthorne and Miller 1949).

The Mathematical Services Department also carried out advisory work. From August 1945 when the Computing Laboratory was first set up, people came to the RAE for advice on punched card machine techniques. One of the first visitors to the RAE Laboratory was Womersley who was collecting information in preparation for punched card machines being installed at the NPL. During the late 1940s, Fairthorne and his staff had a considerable amount of contact with the NPL Mathematics Division concerning punched card machines which developed later into an interest in electronic computers. The RAE also advised the Ministry of Works and the Ministry of Civil Aviation. Therefore, although the RAE Mathematical Services Department was set up to provide centralized computing facilities at the RAE, it had a wider range of users than simply the staff at Farnborough.

The Mathematical Services Department also carried out research on computing machinery. A relay calculator, the RASCAL, was designed at the RAE in the late 1940s (Petherick 1949) but later a decision was made to build the machine using electronic components. Although the design for the RASCAL was complete by 1953, the machine was never finished (Campbell-Kelly 1980*c*). Because of its service work both inside and outside the RAE, the RAE Mathematical Services Department was one of the more well-known, and hence influential, computing centres of the late 1940s and early 1950s.

THE LIVERPOOL MATHEMATICAL LABORATORY

The Liverpool Mathematical Laboratory was set up to serve the staff of the Applied Mathematics Department at Liverpool University, but it too took on outside work although on a much smaller scale than the RAE. Through its computing service work the Liverpool Mathematical Laboratory became reasonably well known as a computing centre.

The Mathematical Laboratory of the Applied Mathematics Department at Liverpool University was set up in 1933 by L. Rosenhead, Professor of Applied Mathematics. It was initially equipped with desk calculating machines supplemented by a collection of mathematical tables. Indeed Comrie's advice on calculating machines and suitable library volumes was sought before the laboratory was set up. The laboratory was intended to be a teaching resource for mathematics students. The Liverpool Mathematical Laboratory was not created as a centralized computing facility for Liverpool University but rather as a laboratory for use by the Applied Mathematics Department alone.

One of the most well-known projects undertaken by the laboratory was the preparation of a working index of mathematical tables by Rosenhead, J. C. P. Miller, and A. Fletcher published as the *Index of mathematical tables* in 1946 by the Scientific Computing Service discussed above. In addition to their work on the *Index* the staff of the Liverpool Mathematical Laboratory computed a considerable number of tables. Miller was an important member of the British Association Mathematical Tables Committee and used the facilities at Liverpool to compute tables in the name of the Committee (Miller 1979). In February 1946 the British Association Mathematical Tables Committee sent the Liverpool Mathematical Laboratory a National Accounting Machine Class 3000 on indefinite loan.[3] Although the machine was used to compute tables for the British Association Mathematical Tables Committee and the Royal Society Mathematical Tables Committee (which took over the British Association's responsibilities for table-making and publishing in 1949), the Liverpool Mathematical Laboratory bore the cost of insurance and maintenance. Over the next eight years the machine was used extensively for Mathematical Table Committee work.

Under the conditions of the loan agreement for the National Accounting Machine between the Mathematical Table Committee and the Liverpool Mathematical Laboratory, the machine was also to be used for teaching and for research into numerical methods. In the early 1950s a certain amount of research work was carried out, particularly by A. Young who joined the Liverpool Mathematical Laboratory in 1951. Young explored the use of the National Accounting machine for integration and for the summation of products (Young 1954). It is interesting to note that the use of Hollerith punched card machines for scientific work was introduced to the Liverpool Mathematical Laboratory by E. Gittus who had joined the Social Science Department at Liverpool in 1951. Gittus had previously worked for the SCS under Comrie and had had considerable experience in handling the X-ray crystallography work done by the company. Although Young experimented with Hollerith machines at the Royal Assurance Company Ltd, punched card machines were not installed at the Liverpool Mathematical Laboratory during the early 1950s.

The Liverpool Mathematical Laboratory was not set up as a computing centre for the university as a whole or as a service centre for outside workers. However, through its work on the *Index of mathematical tables*, its work for the British Association and, later, the Royal Society Mathematical Table Committee, and the work done in applying the National Accounting machine to mathematical problems it became quite well known as a computing centre.

OTHER COMPUTING CENTRES

Apart from the RAE and Liverpool University few of the other computing facilities set up in Britain performed any degree of outside computing work. Some establishments did, none the less, expand or create internal computing facilities. For example the Nautical Almanac Office. In 1945 the NAO was relieved of its Admiralty Computing Service responsibilities. The additional staff which had been assigned to the NAO were transferred to the NPL Mathematics Division. The NAO no longer took on outside work but concentrated on astronomical calculations for its own use. The NAO also continued to develop numerical procedures for use with desk machines and the National Accounting Machine.[4] After the Royal Greenwich Observatory moved to Herstmonceaux Castle in Sussex in 1949, punched card machines were installed at the NAO thereby greatly increasing the computing power of the Office. The installation included an IBM 602A calculating punch, one of the first to be imported into Britain, and much was done to develop its use for the work of the NAO. The NAO thus continued its pre-war role as an establishment which performed a considerable amount of computation, for which it was well equipped, but it did not take on outside work.

Like the NAO several other government establishments installed punched card machines in the late 1940s and early 1950s. For example, the Statistics Department at the Rothampsted Experimental Station at Harpenden installed punched card machines in 1949 at the request of the Ministry of Agriculture and Fisheries. The installation was used primarily for the processing of agricultural surveys but was found to be 'relatively ineffective' as complex calculations were required on small portions of data at frequent and intermittent intervals which necessitated numerous machine interruptions (Yates and Rees 1958). Rothampsted replaced the installation in 1954 with an Elliot 401 computer.

Other establishments had more success with their punched card installations which were sometimes installed as a result of work down on the Hollerith installation at the NPL Mathematics Division. In 1946 the Ministry of Works was one of the first institutions to second staff to the NPL in preparation for its own installation. The two largest punched card installations to result from work carried out by the NPL Punched Card Section were set up at the Atomic Energy Research Establishment at Harwell and the Atomic Weapons Research Establishment at Aldermaston. At Harwell the Hollerith installation was part of the Harwell computing group which was originally set up in 1947 for the use of the Theoretical Physics Division. By 1950 the facility had expanded to provide computing services for the whole of the establishment. At Aldermaston, however, the punched card installation was an isolated facility used only by the Physics Division. Neither Harwell nor Aldermaston took on outside work; this comes as no surprise considering the nature of the research in which they were involved. In any case both installations were heavily used by their own staff. Again both installations were based on experience gained at the NPL which carried out the preliminary research. In both cases the installations were run by staff trained at the NPL. The computational demands at Harwell and Aldermaston were so great that computers were rapidly installed at both installations to first supplement the punched card installations and later replace them.

The replacement of punched card machines by electronic stored program computers was to become a feature of the late 1950s and the 1960s. This is why centres which began to develop computing machinery directly after the war are of such importance.

ENDNOTES TO CHAPTER 9

1 See Chapter 6.

2 Three months after Pugsley had left the RAE to become Professor of Engineering at Merchant Ventures' College, Bristol, the Structural and Mechanical Engineering Department was reorganized as the Structures Department under P. B. Walker.

3 As this machine had been in use by the committee since 1933, it is reasonable to assume that it is the same machine Comrie used on behalf of the Committee before a National was installed at the NAO. See Chapter 3.

4 See, for example, Carter and Sadler 1948.

10

Computing machine developments at Cambridge and Manchester

In Britain in the late 1940s there were three main centres which became heavily, and successfully, involved in the development of electronic stored program computers. These three centres were the NPL Mathematics Division, Cambridge University, and Manchester University. The computer research at the NPL has already been discussed and was part of the work which the Mathematics Division carried out as part of its role as a national computing centre. At Cambridge and Manchester the situation was very different. At Cambridge the existing Mathematical Laboratory was used as a vehicle for computing machinery development. At Manchester the computing centre developed around the research interests of the engineers and mathematicians based at the university.

THE CAMBRIDGE UNIVERSITY MATHEMATICAL LABORATORY

In 1942 John Lennard-Jones, Director of the Cambridge University Mathematical Laboratory, was appointed Chief Superintendent of Armament Research, Ministry of Supply and, although his work took him away from Cambridge, he maintained personal contact with the laboratory and the Ministry of Supply staff working there. During a period of intense post-war planning in February 1945, Lennard-Jones met with the university authorities to discuss the future of the University Mathematical Laboratory. Lennard-Jones perceived that the three main functions of the University Mathematical Laboratory were first, to develop new types of computing machines, secondly, to teach practical computing techniques, and thirdly, to provide a computing service to Cambridge scientists. These functions differed slightly from the original aims of the laboratory which placed a greater emphasis on the provision of a well-equipped computing centre where Cambridge scientists carried out their own computation. The war had, however, revealed both the need for improved computing machines and the need for practical computation to be taught at undergraduate level.

In August 1945 Lennard-Jones took up the post of Director General of Scientific Research (Defence) of the Ministry of Supply and could no longer act as Director of the Mathematical Laboratory. As a result M. V. Wilkes, who had worked in the Mathematical Laboratory before the war, was appointed acting Director of the laboratory and temporary university lecturer in mathematics. Wilkes returned to Cambridge from war service in September 1945 to take up the post. Although the Ministry of Supply had been due to release the Mathematical Laboratory back to the university in September 1945, no alternative accommodation could be found for the Ministry of Supply External Ballistics Group and the laboratory remained occupied until 5 January 1946.

From August 1945 onwards, after his appointment as acting Director had been announced, Wilkes began to plan a reorganization of the Cambridge Mathematical Laboratory. Wilkes contacted both Comrie and Womersley during this time. Comrie wrote to Wilkes in late August stating how he saw the role of the Cambridge Mathematical Laboratory developing (Woodger 1977, item 1). Comrie envisaged a 'triangular team' comprising of the NPL Mathematics Division, the SCS, and the Cambridge University Mathematical Laboratory. In Comrie's opinion, the role of the Cambridge Mathematical Laboratory would be to teach computation to science students and undertake a considerable amount of numerical research. That of the NPL Mathematics Division would be to build computing machinery and develop methods for their use. The SCS would continue to provide a computing service to government research establishments, industry, and the universities. In Comrie's view Cambridge was to become a source of 'computing minded mathematicians' for recruitment to Teddington. To this end Comrie proposed that the SCS would employ some mathematics students during the vacations to give them practical experience.

Wilkes, however, did not foresee the Mathematical Laboratory fitting into the role which Comrie had proposed. Rather, Wilkes saw his first tasks as equipping the laboratory to serve as a university computing centre and planning the future teaching responsibilities of the laboratory. To re-establish the laboratory as a computing centre Wilkes sought the help of Comrie, Womersley, and Fairthorne each of whom had experience in setting up and running computing laboratories. Wilkes visited both Womersley at Teddington and Fairthorne at Farnborough for advice and support. He also received practical help from Comrie who helped Wilkes acquire several desk machines and sold him a second hand National Accounting Machine for the Mathematical Laboratory.

After the Ministry of Supply had released the University Mathematical Laboratory Wilkes prepared a report for the Faculty Board of Mathematics which outlined the work he had done in re-establishing the University Mathematical Laboratory and briefly described his plans for the future (Wilkes 1946*b*). By February 1946 Wilkes had appointed a technician to help him run the machines in the laboratory and was in the process of setting up a workshop which would be used to maintain the differential analyser and the other computing machines. He reported that both the full-size differential analyser and the model machine were in working order but that the Mallock Machine needed extensive rewiring. Wilkes had also acquired two electric Marchant calculating machines, six Brunsvigas, and an electric Victor adding and listing machine. He also had authority to purchase two more electric and five hand calculators once the Board of Trade had issued the necessary licences.[1] Wilkes had also acquired the National Accounting Machine from Comrie and had reserved laboratory space for the installation of a set of Hollerith punched card machines. In addition, with the help of Comrie, Wilkes had established a library of mathematical books and tables and had obtained over 100 titles. The equipment which Wilkes planned to install in the Cambridge University Mathematical Laboratory was very complete and appears to have been modelled on the different sections of the NPL Mathematics Division. The staff of the Cambridge University Mathematical Laboratory was, however, much smaller than that of the NPL Mathematics Division. Apart from Wilkes himself, the only other member of staff was the technician. However Wilkes did have plans to employ an assistant to run the differential analyser and six other staff to operate the calculating and accounting machines.

In his report Wilkes outlined what he saw as the two primary functions of the Cambridge Mathematical Laboratory. The first of these functions was to provide a course on practical computation for all science undergraduates at Cambridge. Although Wilkes had not, in early 1946, planned such a course in any detail nor employed a lecturer to run it, he proposed that the 'course should be thoroughly practical and should aim at giving sound advice on how to tackle the ordinary problems likely to be met with, and how to make the best use of modern calculating machines and aids to computation' (Wilkes 1946*b*, p.2).

The second important function of the Mathematical Laboratory was to provide a computing facility for Cambridge research staff. Wilkes had already begun to establish a well-equipped laboratory which research workers could use to carry out their own work using desk calculators or the Mallock Machine. The staff of desk machine operators which Wilkes proposed to employ would be on hand to offer advice and assistance to workers using the laboratory. The computing staff would only take on large repetitive computing jobs for research workers after the problem had been reduced to a series of routine steps. Wilkes was insistent that any work performed by the laboratory staff would have to be supervised by the research worker involved and not by Wilkes or senior Mathematical Laboratory staff. The Mathematical Laboratory did not, therefore, propose to provide a complete computing service in the same sense as the SCS or the NPL Mathematics Division, but rather Wilkes provided workers with the facilities to carry out their own computational work.

The use of the differential analyser, however, fell into a different category because experienced staff were needed to operate the machine properly. Work which required the use of the differential analyser would, therefore, have been carried out by laboratory staff. Wilkes envisaged that those wishing to use the model differential analyser, particularly research students, would want to operate the machine themselves and he had no objection to this.

In his 1946 report Wilkes stressed that the facilities provided by the University Mathematical Laboratory, with the exception of the Mallock Machine, should not be made available to workers outside Cambridge University. There were two reasons for this. First, Wilkes predicted that the laboratory would be used to capacity by university staff and students alone. Secondly the system of individuals supervising and performing their own work was unworkable if the user was not resident in Cambridge. Because the Mallock Machine was the only one of its kind in the country, Wilkes proposed that this machine alone be made available, for a fee, to people outside the university.

While the provision of a well-equipped computing laboratory and the laboratory's teaching role are the two functions emphasized by his report, it is clear that Wilkes also perceived the laboratory undertaking two different types of research. Firstly, research into numerical procedures for use with calculating machines which would arise naturally from the various computations carried out by the laboratory. Secondly, Wilkes suggested that the University Mathematical Laboratory should carry out research into new types of electronic calculating machines and try to 'catch up some of the lead which the Americans have in the subject' (Wilkes 1946*b*, p.6). The development of electronic calculating machines was where Wilkes's main interests lay.

As a result of this report, and a statement from Lennard-Jones indicating that he did not wish to continue as Director of the Mathematical Laboratory when he returned to Cambridge, Wilkes was appointed Director of the laboratory for a period of five years from October 1946. At the same time a Mathematical Laboratory Committee consisting of members from all interested university faculties was set up. Wilkes was directly responsible to this committee.

During 1946 Wilkes's ideas on the type of computing machinery research which the Mathematical Laboratory was to carry out began to crystallize. In 1945 and 1946 D. R. Hartree (who was appointed to the Plummer Chair of Mathematical Physics at Cambridge in 1946) visited the Moore School of Electrical Engineering at the University of Pennsylvania, Philadelphia to work on and to advise on the ENIAC machine. Hartree told Wilkes of the work being done at the Moore School. Comrie too visited the United States in the post-war period and, on his return, lent Wilkes a copy of von Neumann's 'First draft of a report of the EDVAC'.[2] In the summer of 1946 Hartree arranged for Wilkes to attend the now famous Moore School lectures where Eckert, Mauchly, Goldstine, von Neumann, Burks, and others lectured on their work with electronic computing machines including the proposed EDVAC machine (Chambers 1947).

In October 1946 Hartree delivered his inaugural lecture as Professor of Mathematical Physics. His lecture had the title Calculating machines: recent and prospective developments and their impact on mathematical physics. In his lecture Hartree warned physicists and others to 'begin looking ahead and thinking what they would like to do when equipment of the kind I am going to talk about becomes more generally available then at present' (Hartree 1947*b*, p.7). Hartree describes the ENIAC, but more importantly also discussed the basic principle of the stored program computer. By delivering such a lecture just as Wilkes was beginning his research on electronic computers Hartree was setting the stage for Cambridge to become an important *computer* centre rather than a computing centre.

Wilkes returned from the United States bursting with ideas about building an EDVAC type machine. Hartree, who sat on the NPL Executive Committee and who had special responsibility for the work of the Mathematics Division, encouraged Wilkes to approach Womersley with a view to collaborating on the construction of such a machine. At that time, autumn 1946, Turing was working on the design of the ACE machine at Teddington. As the Mathemetics Division did not intend, at that time, to build the machine at Teddington, Hartree saw the possibility of collaboration between the NPL and the University Mathematical Laboratory. Wilkes prepared a proposal for the construction of a prototype machine based on the EDVAC (Woodger 1977, item 9). Turing disagreed with the suggestions Wilkes put foward and told Womersley that they were 'very contrary to the line of development here [at Teddington], and much more in the American tradition' (Woodger 1977). Despite Turing's disapproval of Wilke's proposed machine, Womersley continued to consider the possibility of asking Cambridge to build a machine on behalf of the NPL. After months of delay during which F. C. Williams, from the Telecommunications Research Establishment and later Manchester University, turned down a contract with the NPL, the NPL decided to construct a machine themselves. In April 1947 Womersley wrote to Wilkes stating quite clearly that no such contract was to be made (Woodger 1977). In the meantime, however, Wilkes had begun to construct his own machine—the EDSAC.

In July 1947 the Mathematical Laboratory Committee prepared a progress report which it presented to the Faculty Board of Mathematics (University of Cambridge, Mathematical Laboratory Committee 1947). The report stated that, besides Wilkes, the staff of the Mathematical Laboratory now consisted of a research assistant working on the differential analyser with two 'boys' to assist him, a technician, a mathematician, one secretary, and three desk machine staff. Wilkes found it impossible to find suitable computing staff and never took on the six he had earlier proposed. The report noted that the laboratory was equipped with 12 desk calculators, a National Accounting Machine, a Victor adding and listing machine, the model and full-size differential analysers, and the Mallock Machine, which had been extensively rewired. A set of Hollerith punched card machines on loan from BTM were also to be installed in the laboratory. These machines had previously been used by S. Chapman at Imperial College for tidal motion calculations. When Chapman moved to Oxford in 1946 he suggested that Wilkes, who had an interest in the work, take over the calculation and the machines. Wilkes had always intended to install punched card machines in the laboratory (Wilkes 1946*b*) and gratefully accepted the offer. Although the Hollerith machines were used to continue Chapman's work their principle use at Cambridge was for X-ray crystallography work by a team from the Cavendish Laboratory led by W. Cochran (Wheeler 1983*a*; Hodgson *et al.* 1949).

Wilkes's 1947 progress report listed five major functions of the Cambridge University Mathematical Laboratory. These were

(i) Research in computational methods and on all kinds of calculating machines.

(ii) The carrying out of computational work by the Laboratory staff for research workers in all faculties.

(iii) The provision, in co-operation with the Faculty Board, of teaching on computation and on the construction and use of calculating machines.

(iv) The provision of advice on all questions concerning computation.

(v) The provision of computing rooms and machines for the use of individual research workers.

(University of Cambridge, Mathematical Laboratory Committee 1947, p.4).

It is interesting to note how the emphasis placed on the different aspects of the laboratory's work in the 1947 report differed from those outlined in Wilkes's report a year previously (Wilkes 1946*b*). In the earlier report Wilkes had clearly stated that the two most important functions of the Mathematical Laboratory were to provide computing facilities for university staff and to run courses for undergraduates. By 1947, however, the main emphasis had shifted towards computing machinery research. Although the Cambridge Mathematical Laboratory provided both computing facilities for Cambridge scientists and courses on computational mathematics,[3] the main work of the laboratory was the design and construction of an electronic computer called EDSAC. With the help of finance from the Department of Scientific and Industrial Research, the University Grants Committee, and J. Lyons & Company (which also seconded staff to Cambridge) the EDSAC was rapidly built. It ran its first automatic calculation in May 1949 and went into regular use early in 1950[4] (Plate 11).

Although the University Mathematical Laboratory was always first and foremost for the use of Cambridge staff and students, the laboratory soon became an important national computing centre. In 1947 a series of Thursday afternoon seminars began, which ran for several years, and attracted people from all over Britain who were interested in electronic computers. In June 1949 Cambridge hosted a computer conference which was attended by nearly all those working in the field (University Mathematics Laboratory, Cambridge 1950). The Mathematical Laboratory was also an important training ground not only for Cambridge students, but for other institutions which seconded staff to Cambridge, for example the Atomic Weapons Research Establishment at Aldermaston sent A. E. Glennie and K. N. Dodd to Cambridge, and P. Taylor was sent from the Telecommunications Research Establishment.

After the EDSAC went into regular service in 1950, the Mathematical Laboratory began to take in work from outside Cambridge. To deal with the large number of Mathematical Laboratory users a priorities committee was set up to allocate time on the EDSAC and to advise users. Because programmers were relatively few most users had to do their own programming and in this way the Mathematical Laboratory also became a training ground for many early computer users. Consequently during the 1950s the Cambridge Mathematical Laboratory became an important national computer centre chiefly because of its research into computing machinery and programming systems, and the computing service it subsequently offered.

THE ROYAL SOCIETY COMPUTING MACHINE LABORATORY AT MANCHESTER UNIVERSITY

Although a significant computer centre had been built up around Hartree's differential analyser at Manchester University before and during the Second World War, the later research on electronic computers at Manchester was virtually unconnected with the previous differential analyser work.

In October 1945 M. H. A. Newman took up the Chair of Pure Mathematics at Manchester University. During the war Newman had been based at the Government Code and Cypher School, Bletchley Park. There he had been involved in the wartime development of the Colossi and Heath Robinson machines which were built in conjunction with staff from the Post Office Research Station at Dollis Hill.[5] Through this work and his acquaintance with Turing, who had also been at Bletchley, Newman was aware of some of the work being done on electronic computers. One of Newman's reasons for accepting the post at Manchester was that he hoped to be able to set up a computing laboratory at the university in order to study advanced pure mathematical problems using computers. P. M. S. Blackett, Professor of Physics at Manchester University, had held several important scientific and administrative posts during the war and was aware of the work at Bletchley, and elsewhere, concerning computers. Blackett too was keen to see a computer laboratory set up at Manchester and encouraged Newman to apply for a Royal Society grant to fund such a project.

In February 1946 Newman applied to the Royal Society for a grant to set up a computer laboratory at Manchester. Newman proposed that, rather than working on the design and construction of electronic computers, his group in the Mathematics Department would concentrate on the 'mathematical and logical problems of finding the best use of such machines and investigate their effect on the development of mathematics (Newman 1948) Newman wrote

> with the development of fast machine techniques mathematical analysis itself may take a new slant, apart from the developments that may be stimulated in symbolic logic and other topics not usually in the repertoire of engineers of computing experts; and that mathematical problems of an entirely different kind from those so far tackled by machines might be tried, e.g. to test out, the 4-colour problem or various theorems on lattices, groups, etc., for the first few values of n. (Newman 1946)

Newman intended to use the money awarded to buy in the components necessary to build a computer as soon as sufficient engineering progress had been made towards the construction of a working machine.

The Royal Society set up a committee to consider Newman's application for a grant to establish a computer laboratory at Manchester University. The committee membership comprised Blackett, who already supported the project, Hartree, influential in the computer projects at Cambridge and the NPL, Darwin, Director of the NPL, W. V. D. Hodge, a pure mathematician from Cambridge, and J. H.C. Whitehead, a pure mathematician from Oxford. Darwin, both as Director of the NPL and an instigator of the Mathematics Division, was strongly opposed to awarding such a grant. He felt that the creation of a computer laboratory at Manchester would undermine the NPL's ACE project. When completed Darwin expected the ACE to serve the computing needs of the whole country. Darwin was also concerned that Newman would enlist the help of T. H. Flowers to assist him (Hodges 1983). Flowers, an old colleague of Newman's from Bletchley, was then working at the Post Office Research Station at Dollis Hill which had been commissioned by the NPL to develop delay lines for the ACE. Despite Darwin's objections the committee granted Newman £35 000 in June 1946 for the establishment of a Computing Machine Laboratory at Manchester (Royal Society 1946, 13 June). The Royal Society awarded the grant on the understanding that the machine Newman proposed to build at Manchester would not duplicate the work of the NPL; the understanding was that the ACE was to be used for numerical computations while any machine at Manchester was to be applied to pure mathematics. The Royal Society stipulated that £20 000 of the grant was to be spent on the construction of a suitable machine while the remainder was to be allocated to staff salaries over a 5 year period.

Newman received the Royal Society grant in summer 1946 and immediately used it to send D. Rees, whom Newman had brought with him from Bletchley, to the United States to attend the Moore School lectures in July and

August 1946. Other than that there was little Newman could do until someone, either in Britain or the United States, constructed a working machine or a working memory unit. Newman was a mathematician not an engineer and had no intention of trying to design and build a machine himself. Initially Newman anticipated that the first machine which would be available to him would arise from John von Neumann's work at Princeton. However computing machine research unexpectedly came to Manchester.

The development of computer hardware at Manchester University sprang from the work of F. C. Williams. Before the war Williams had been an assistant lecturer at Manchester University and had worked with Blackett in designing the automatic curve follower for Hartree's differential analyser. During the war, Williams had been heavily involved in the development of radar at Bawdsey and later at the Ministry of Supply Telecommunication Research Establishment, Malvern (TRE). During two trips to the Radiation Laboratory at the Massachusetts Institute of Technology (MIT), the first in November 1945 and the second in June 1946, Williams learnt about the work being done in trying to use cathode ray tubes (CRTs) as storage devices. This application of CRTs greatly interested Williams and he began to carry out experiments of his own.

In May 1946 Darwin and Womersley approached Williams and the staff at the TRE and asked them to undertake development of a delay line store for the ACE project because work on delay lines for the NPL at Dollis Hill was proceeding too slowly. This request did not coincide with William's interest in CRTs and Williams spent more time developing CRTs than looking at delay lines. In December 1946 Williams left the TRE to become Professor of Electrical Engineering at Manchester University. Around the same time several TRE staff transferred to the Department of Atomic Energy. As a result the TRE no longer felt that it had the expertise to undertake the type of work which the NPL required. Darwin and Womersley were, however, still hopeful that Williams would carry out the necessary work. Consequently they approached Williams directly and offered him a contract under which Williams would construct a storage device and other components for the NPL ACE to his own design. Hodges (1983) points out the absurdity of asking Williams to develop a CRT store for the NPL when Turing's design for the ACE was based around the use of delay lines. Hodges remarks that 'Very probably Darwin and Womersley did not appreciate that the storage medium dictated many aspects of the design of the machine and its programming' (Hodges 1983, p.350). Williams turned down the contract with NPL preferring instead to work independently on CRT storage devices.

Despite Williams's move to Manchester the TRE continued to support Williams's research by supplying components to Manchester which would have been extremely difficult to obtain elsewhere (Lavington 1975). Williams also arranged for his assistant Tom Kilburn to be seconded to Manchester in order to complete his Ph.D. Williams's work on computing devices at Manchester was not, therefore, funded by Newman's Royal Society grant. Indeed Newman's only contributions to William's work over the following two years, apart from moral support, was the supply of war surplus equipment from Bletchley and a brief explanation to Williams of the principle of storing numbers and instructions. Williams recalled

> I never was, never have been, and never will be a mathematician. I did not even know that there was any system of numbers other than the scale of ten, but when the specification of a storage system was explained to me I could grasp what was wanted. (Williams 1975, p.327)

Development of the CRT store proceeded quickly and by autumn 1947 the Williams tube was able to store 2048 bits. Following this a 'baby' prototype machine was built to test the store, and first ran successfully on 21 June 1948. Although very small, the memory had only 32 words of 32 bits each, the machine was an important one. Not only was the machine the world's first working stored program computer and proved that a computer could be built around a CRT memory, but it also won the necessary finance for Williams to continue his development work independently of Newman's Royal Society grant.

Soon after the Manchester 'baby' machine had been demonstrated Blackett suggested that it was an invention from which industry would benefit (Drath 1973). Despite Williams feeling that this suggestion was premature Blackett contacted Ben Lockspeiser, Chief Scientist Ministry of Supply and described the work at Manchester. Lockspeiser visited Manchester to see the machine and was sufficiently impressed to arrange a meeting between Williams and Ferranti Ltd to discuss a joint venture to build an engineered full-size computer with Ministry of Supply funding.[6]

The Ministry of Supply was interested in funding computer research for three reasons: first, to benefit scientific research generally; second, to benefit defence-related research; third, to promote industrial development. The third reason was an important one. In the late 1940s much thought was given to post-war planning and the revival of British industry. Part of the role of the Ministry of Supply was to encourage the commercial application of basic research with the help of Treasury funding. The Ministry of Supply felt that simply making the appropriate patents available[7] would not be sufficient to stimulate commercial interest in computers. Funding was needed. Consequently in autumn 1948 Lockspeiser organized a £120 000 grant for Ferranti to cover the cost of building a computer to Williams's design. The computer built was to be installed at the university with the maintenance costs being met by the Ministry of Supply.

Because of the money put forward by the Ministry of Supply, Williams again had little need of Newman's Royal Society grant. The Royal Society

Computing Machine Laboratory at Manchester University consisted only of a single room which housed the evolving prototype machine now known as the Manchester Mark I (Plate 12). Throughout 1948 and 1949 Williams, with his growing team in the Electrical Engineering Department which by now included D. G. B. Edwards, G. C. Tootill, A. A. Robertson, and G. E. Thomas, continued to develop the machine. In late 1949 the design was sufficiently finalized to be handed over to Ferranti.

As the Manchester Mark I developed into what was obviously going to be a powerful computing tool, Newman started to develop the Royal Society Computing Machine Laboratory by establishing the programmer and user support which would be needed once the machine went into operation. As part of this process Newman asked the university to create a readership for Alan Turing, who was spending a sabbatical year from the NPL at Kings College, Cambridge. Turing's salary was to be paid from the Royal Society grant. Turing, disillusioned with the ACE project at the NPL, accepted the post. Although Turing did not take up the appointment until September 1948, he wrote to Williams during the preceding months asking for machine details so that he could begin to write programs for the Manchester machine.

Although Turing held the nominal title of Deputy Director of the Royal Society Computing Machine Laboratory under Newman, neither Newman nor Turing had any influence on Williams's designs for the computer. Newman worked on developing mathematical problems for the computer until his interest dwindled towards the end of 1949. With Tootill and Edwards, Turing developed a programming system for the prototype Mark I machine using teleprinter codes. He also prepared a programmers manual and began to write programs for the machine.

In preparation for the arrival of the engineered version of the Mark I from Ferranti, the staff of the Royal Society Computing Machine Laboratory increased in October 1949 with the appointment of C. Popplewell as Mathematical Assistant. A. Bates, a postgraduate student working under Turing, also arrived around this time. Both women at first shared an office with Turing and both contributed to the programming work. The construction of a new building designed to house the Ferranti machine was also paid for by the Royal Society grant. The building was completed in January 1951 and the staff, which expanded in mid-1951 to include R. K. Livesley and N. E. Hoskin, quickly moved in. The Ferranti Mark I was installed in February 1951. To coincide with the inauguration of the Ferranti Mark I, in July 1951, Manchester hosted the second British Computer Conference. Like the 1949 conference at Cambridge nearly all those working in the field in Britain attended. At about this time Turing's interests turned from computers to morphogenesis but, although Turing was no longer directly involved in the work of the Computing Machine Laboratory, he continued to be an important machine user.

The Ferranti Mark I provided much more computer power than the relatively small Computing Machine Laboratory required in 1951 and, consequently, outside users came to use the machine. The computer service run by the university was supported by Computing Machine Laboratory staff who both wrote system software and assisted, advised, and trained outside users who were charged £20 per hour to use the machine. Some of the income generated by offering a computer service to outside users was used to fund further machine development at Manchester. The remainder helped to keep the service in operation supplemented by the university and the Department of Scientific and Industrial Research (DSIR) which had taken over the Ministry of Supply's responsibilities concerning the machine. Machine development was the most important aspect of the work at Manchester University. Provision of a computing service was not the main interest of the individuals involved. This contrasts sharply with the NPL Mathematics Division where the original aim was to provide a national computing service. At the Cambridge Mathematical Laboratory the initial aim had been only to provide local computing facilities but when the EDSAC went into regular operation its services were offered to outside users.

A. E. Glennie, an early user at both Manchester and Cambridge, recalls a different atmosphere in the two centres (Glennie 1978). He found that while there was considerable freedom at Manchester to experiment with different programming methods, there was very little in the way of user support. At Cambridge there was plenty of user support but access to the machine was restricted and the programming methods used were well established. The NPL Mathematics Division and the Cambridge Mathematical Laboratory saw themselves as computing service centres but Manchester did not adopt this attitude quickly. From the outset Williams's interest had been an academic one of trying to devise a working machine. Newman too had not intended to run a service but to use the machine for mathematical research. Thus the work done at Manchester during the late 1940s was not aimed at providing a user-friendly computing service. The early programming system, for example, was based on teleprinter codes which were not easy for the uninitiated to grasp.

Although Williams's interests soon turned from computers to other aspects of electrical engineering, computer development continued at Manchester under Tom Kilburn. Over the 20 years following the installation of the Ferranti Mark I in 1951 the collaboration between Manchester University and Ferranti produced four other commercially available computers. The Mercury, the Metropolitan-Vickers MV 950, the Atlas, and the ICL 2980. Thus Manchester University continued to be a centre for the development of computer hardware. Manchester was also responsible for several significant software developments, particularly the Autocode which R. A. Brooker developed for the Mercury machine. These innovations and the range of users which came to Manchester to work on the machine (including A. E. Glennie, K. N. Dodd, and C. Strachey) combined to make Manchester a very important computer centre.

CONCLUDING REMARKS

Following the success of the computer projects at Cambridge, Manchester, and the NPL, Lyons, Ferranti, English Electric, Elliot Brothers, and others began to commercially manufacture computers. American companies too, most notably IBM, started to produce electronic stored program computers in line with research taking place in the United States. By the late 1960s computers were an important part of scientific research establishments and universities. The need for a national computing centre gradually, and naturally, declined. Computation began to be decentralized; first to the level of one machine to a single university or research establishment, and now, today, to the level of the individual possessing a powerful personal computer which may, or may not, be linked into a larger shared machine.

In studying the gradual centralization of computing power during the first half of the twentieth century it has been shown that centralization did not automatically follow advances in technology. The development of desk calculating machines in the late 1800s made it theoretically possible to centralize some types of calculation. However it was not until Comrie's work on numerical methods for use with desk machines in the late 1920s that some kind of centralized computing became a reality. The installation of the differential analyser at Manchester University did not prompt the creation of a computing centre in the sense described in Chapter 1. Rather the machine formed the hub of a changing group of people who wished to use the machine for a certain class of problems but who could not afford the expense of purchasing a differential analyser themselves.

The advantages of centralizing scientific computation were highlighted during the Second World War when the need to carry out mathematical work quickly and efficiently was felt in many quarters. Yet it was not until 1942 when the Admiralty Computing Service was set up that a centralized computing facility was created within one of the service ministries. However the widespread sense of a need to centralize computing resources, which the war brought about, made possible the creation of a national computing centre in the form of the NPL Mathematics Division.

The most important single factor which prompted the creation of computing centres was the work of far-sighted individuals: Pearson introducing desk calculating machines to his laboratory; Comrie revolutionizing the NAO; Comrie setting up the SCS; Hartree importing the differential analyser; Lennard-Jones creating a computing centre at a university; Todd, Sadler, and Carroll pressing for the ACS, and later the NPL Mathematics Division, to be set up. In all cases centralized computing facilities were created when individual scientists and mathematicians perceived their need and when those individuals were in a politically strong enough position to carry their ideas through to reality.

The work of individual computing centres also encouraged and made possible further centralizations. For example, Comrie was influenced by Karl Pearson's work on desk calculating machines; Comrie's experience at the NAO drew his attention to the need for a computing service; Lennard-Jones argued his case for a computing laboratory on the basis of both Hartree's and Comrie's work; the ACS drew on the experience of the NAO; and the NPL Mathematics Division was heavily based on the ACS. Each computing centre used the work and experiences of computing centres before them and built up from there.

After the NPL Mathematics Division had been set up the pattern changed. Local computing centres such as those at the RAE and the AERE, Harwell, were set up to provide local computing facilities but kept in close contact with the NPL. Others, like the Liverpool Mathematical Laboratory, were specialist centres. Once computers became a practical reality and more and more people had become familiar with their operation through time spent at the NPL, Cambridge, or Manchester, centralization on a national scale became history.

END NOTES TO CHAPTER 10

1 Immediately after the war calculating machines of all kinds were in very short supply and permission was needed from the Board of Trade before any machine could be purchased.

2 Although only intended for limited distribution this document was widely circulated. It was the first written account of the stored program computer and greatly influenced future computer projects.

3 The most important of these lecture courses was the annually delivered numerical analysis lectures given by Hartree. The numerical analysis lectures began in 1947 and resulted in Hartree's book of the same name which was published in 1952.

4 Details of the logical design and programming systems of EDSAC have been discussed by others. See for example Campbell-Kelly 1980*a*; Campbell-Kelly 1982; University Mathematics Laboratory, Cambridge 1950; Wilkes 1975; Wilkes and Renwick 1949.

5 Although much work was done on electronic computers at Bletchley during the war, much of which has only recently come to light, it was in no sense a scientific computing centre. The work carried out was concerned principally with code-breaking and large-scale data processing. In addition this work was highly classified and its facilities were not available to outsiders. Because Bletchley was not a scientific computing centre in the sense used here, a description of the work it performed during the war has not been included.

6 Manchester University traditionally had strong links with local industry. Ferranti was based in the Manchester area and was suffering from a decline in business

following the end of the war during which it had had numerous government contracts. For example, Williams himself had collaborated with Ferranti during the war on developing a radar device for the identification of friendly aircraft.

7 Because much of Williams's work on CRT stores had been carried out with the support of the TRE the Ministry of Supply held many of the important patents.

APPENDIX 1

APPENDIX 1.1: *Heliocentric places of Mars: last digits of interpolates*

1923 Jan 13 (31 21 51) .31

0

5

5

9

5

3

2

0

6

9

8

1923 Jan 25 (38 20 18) .53

APPENDIX 1.2: *Heliocentric places of Mars: last digits of 1st and 2nd differences of the interpolates*

1923 Jan 13 (31 21 51) .31			
		9	
	0		4
		5	
	5		5
		0	
	5		6
		4	
	9		8
		6	
	5		8
		8	
	3		9
		9	
	2		1
		8	
	0		2
		6	
	6		3
		3	
	9		4
		9	
	8		4
		5	
1923 Jan 25 (38 20 18) .53			

APPENDIX 1.3: *Heliocentric places of Mars: inferred 2nd differences*

	2nd diff.	1st diff.
		(35 13.51)
1923 Jan 13 (31 21 51) .31	(3.22)	
9		
0 4	3.24	
5		
5 5	3.25	
0		
5 6	3.26	
4		
9 8	3.28	
6		
5 8	3.28	
8		
3 9	3.29	
9		
2 1	3.31	
8		
0 2	3.32	
6		
6 3	3.33	
3		
9 4	3.34	
9		
8 4	3.34	
5		
1923 Jan 25 (38 20 18) .53		

APPENDIX 1.4: *Heliocentric places of Mars: 1st differences computed using Burroughs adding and listing machine*

	2nd diff.	1st diff.
		(35 13.51)
1923 Jan 13 (31 21 51) .31	(3.22)	
9		35 10.29
0 4	3.24	
5		35 07.05
5 5	3.25	
0		35 03.80
5 6	3.26	
4		35 00.54
9 8	3.28	
6		34 57.26
5 8	3.28	
8		34 53.98
3 9	3.29	
9		34 50.69
2 1	3.31	
8		34 47.38
0 2	3.32	
6		34 44.06
6 3	3.33	
3		34 40.73
9 4	3.34	
9		34 37.39
8 4	3.34	
5		34 34.05
1923 Jan 25 (38 20 18) .53		

APPENDIX 1.5: *Heliocentric places of Mars: function values computed using the Burroughs Adding and Listing Machine*

(Bold figures indicate overprinting of the 1st differences)

	2nd diff.	1st diff.	Longitude of Mars
		(35 13.51)	
1923 Jan 13 (31 21 51) .31	(3.22)		31 21 51.31 Jan 13
9		**35 10.29**	
0 4	3.24		31 57 01.60 Jan 14
5		**35 07.05**	
5 5	3.25		32 32 08.65 Jan 15
0		**35 03.80**	
5 6	3.26		33 07 12.45 Jan 16
4		**35 00.54**	
9 8	3.28		33 42 12.99 Jan 17
6		**34 57.26**	
5 8	3.28		34 17 10.25 Jan 18
8		**34 53.98**	
3 9	3.29		34 52 04.23 Jan 19
9		**34 50.69**	
2 1	3.31		35 26 54.92 Jan 20
8		**34 47.38**	
0 2	3.32		36 01 42.30 Jan 21
6		**34 44.06**	
6 3	3.33		36 36 26.36 Jan 22
3		**34 40.73**	
9 4	3.34		37 11 07.09 Jan 23
9		**34 37.39**	
8 4	3.34		37 45 44.18 Jan 24
5		**34 34.05**	
1923 Jan 25 (38 20 18) .53			38 20 18.53 Jan 25

APPENDIX 2

Incomplete list of SRE/ACS reports

Report number	Title
7	Summation of certain slowly convergent series
8	Mechanical quadrature
9	Table of $f(x, y) = (2\pi)^{-1} \int_0^{2\Pi} e^{-x.\cos\theta - y.\cos 2\theta}\, d\theta$
18	Cable tables
19	Tables of the Incomplete Hankel functions
20	Trajectory of a body moving with resistance prohibited
21	Calculations involving the Airy Integral for complex arguments first
22	Tables of integrals $A(x) = \int_0^x \frac{\cos(\pi t/2x)}{1+t^2}\, . \, dt$ $B(x) = \int_0^x \frac{\cos(zt/x)}{1+t^2} \cdot dt$; and $C(x) = \int_0^x \frac{\cosh(zt/x)}{1+t^2} \cdot dt$
26	Asymptotic expansions
31	Calculations involving the Airy Integral for complex arguments-second
35	Rapid calculation of bearing tables
37	An electronic differential analyser by J. M. Jackson. Reprinted from a US Navy publication.
39	Calculations involving the Airy Integral-third
40	Rangefinder performance computer
46	Calculations involving the Airy Integral-fourth
47	Tabulation of the function $f(x, y) = \int_0^x e^{-k}\left[\frac{J_0(kx)\cosh(ky)}{\sinh k} - 1\right] . \, dk$
51	Tables of angular quarter squares
52	Tables of certain integrals
53	Dictionary of Laplace transforms. Part 1

55	Calculations involving the Airy Integral for complex arguments-fifth
62	Tables of the function $f(x)=e^{-x}\int_0^{\Pi/2} e^{x\cos\theta}\sin^2\theta\,.\,d\theta$
65	Tables of Legendre functions
68	Dictionary of Laplace transforms. Part 2A
71	Dictionary of Laplace transforms. Part 2B
76	A statistical investigation into night vision of watches at sea
80	Probability charts for destructive tests
82	Zeros of Laguerre polynomials
89	Solution of integral equations occurring in an aerodynamical problem (supersonic flow)
90	Tables of integrals $C(t)=k^{1/2}\int_0^x e^{-u}\cos(2tk^{1/2}u-ku^2)\,du$ $S(t)=k^{1/2}\int_0^x e^{-u}\sin(2tk^{1/2}u-ku^2)\,du$
91	Miscellaneous information sheet III
92	Lateral vibration of beams of conics
93	Tables for the summation of triangulations
95	Stresses in turbine rotors
96	Computation of an integral occurring in the theory of water waves
97	Tables of the Incomplete Airy Integral
101	Dictionary of conformal representations Part I
102	Dictionary of Laplace transforms Part 3A
106	Asymptotic expansions (update of SRE/ACS 26)
107	Dictionary of conformal representations. Part II
108	Dictionary of Laplace transforms. Part 3B
109	Dictionary of conformal representations. Part III
110	Dictionary of conformal representations. Part IV
111	Dictionary of conformal representations. Part V
112	Alan Baxter (1910–1947). The Fourier transformer

APPENDIX 3

APPENDIX 3A: *Provisional research programme for the proposed Central Mathematics Station. 10 May 1944*

1. The development of methods for the solution of large numbers of simultaneous equations, such as arise in triangulations and in Southwell's relaxation process.
2. The application of automatic telephone equipment to computing.
3. The development of methods and/or instruments for the solution of integral equations.
4. The development of methods of finding the characteristic values of Lagrangian systems.
5. The development of methods and machines for the Fourier analysis of both periodic and aperiodic curves, and for evaluating inverse Laplace transformations.
6. Numerical and mechanical solutions of partial differential equations.
7. The development of instruments for the direct solution of vibration equations with given conditions without the use of the Differential Analyser.
8. The development of new methods of statistical planning of industrial experiments for maximum economy and efficiency. Application of these methods to problems of testing and standardisation.
9. The collation and analysis of accumulated results of Quality Control with a view to discovering general principles, such as the performance of different types of machine tools and the nature of the frequency distributions which arise in manufacturing processes.
10. The statistical formulation of problems of design and assembly.
11. The improvement of Quality Control methods, and of methods of Acceptance Sampling Inspection.
12. The invention of new instruments for convenient use in industrial firms and research establishments for their statistical work.

(From the Report of Interdepartmental Technical Committee on a Proposed Central Mathematical Section, Appendix B.)

APPENDIX 3B: *Analysis of subjects covered by the research programme for the proposed Central Mathematical Station. 10 May 1944*

Research project	Statistical	General	Specific	Machine development	Numerical methods
1			X		X
2		X		X	
3			X	X	X
4			X		X
5			X	X	X
6			X	X	X
7			X		X
8	X		X		X
9	X		X		X
10	X		X		X
11	X		X		X
12	X	X		X	
Total	5	2	10	5	10

APPENDIX 4

APPENDIX 4A: *NPL Mathematics Division users 1946–1951*

Admiralty
Armament Research Establishment
Atomic Energy Research Establishment, Harwell
Australian National Standards Laboratory
Bank of England
Birmingham University
Board of Trade
British Cotton Industries Research Association
British Electricity Authority
British Iron & Steel Research Association
British Railways
British Standards Institute
Building Research Station, DSIR
Civil Service Commission
Colonial Survey Department
DSIR Headquarters
Electrical Research Association
'F' Division, DSIR
Fuel Research Station, DSIR
Home Office
Intelligence Division, DSIR
Interdepartmental Committee on Servo Mechanisms
London Passenger Transport
London University
Manchester University
Mechanical Engineering Research Organisation, DSIR
Medical Research Council
Ministry of Agriculture
Ministry of Aircraft Production
Ministry of Civil Aviation
Ministry of Education
Ministry of Health
Ministry of Supply
Ministry of Works
NPL Aerodynamics Division
NPL Metallurgy Division
NPL Metrology Division

NPL Physics Division
NPL Ship Division
Ordnance Survey
Oxford University Crystallography Group
Road Research Laboratory, DSIR
Royal Aircraft Establishment, Farnborough
Royal Society Mathematical Tables Committee
Sir Edward Appleton
Swedish Government Computing Laboratory
Sperry Gyroscope Company Ltd.
Treasury Training Division
United Steel Companies Ltd.
University College London, Statistics Department
War Office

APPENDIX 4B: *NPL Mathematics Division users 1952–1956*

Admiralty
Aircraft Manufacturers
Atomic Energy Research Establishment, Harwell
Atomic Weapons Research Establishment, Aldermaston
BBC and Nuffield Foundation
Cambridge University
Central Electricity Authority
Chemical Research Laboratory
DSIR Headquarters
English Electric
European Council for Nuclear Research
Home Office
Inland Revenue
Institute of Oceanography
London Hospital Medical College
Mechanical Engineering Research Station, DSIR
Ministry of Civil Aviation
Ministry of Supply
National Blood Transfusion Service
National Coal Board
Norwegian Defence Research Department
NPL Aerodynamics Division
NPL CME Division
NPL Director
NPL Electricity Division
NPL Light Division

NPL Metrology Division
NPL Physics Division
NPL Ship Division
Ordnance Survey
Oxford University
Post Office
Road Research Laboratory
Royal Aircraft Establishment, Farnborough
Royal Dutch Shell Company
Royal Society Mathematical Tables Committee (RSMTC)
Standard Telephones & Cables
Torry Research Station, DSIR
United Kingdom Atomic Energy Authority
United States Air Force
University College London, Physics Department
University of Singapore
Water Pollution Research Laboratory, DSIR

References

Aeronautical Research Committee (1937–40). *Minutes of the ARC Oscillation Sub-Committee*, 9th meeting 21/5/37, 18th meeting 9/2/40; 19th meeting 4/10/40. Public Records Office DSIR 22/97, London.

Andoyer, H. (1911). *Nouvelles tables trigonométriques fondamentales (logarithmes)*. Hermann, Paris.

Anon. (1905). From an Oxford notebook. *Observatory*, 361–6.

Anon. (1933). An electrical calculating machine. *Nature*, **132**, 880.

Anon. (1934). The Mallock electrical calculating machine. *Engineering*, **135**, 26 May, 566–8.

Anon. (1935*a*). The differential analyser at Manchester University. *Engineering*, 26 July, 88–90.

Anon. (1935*b*). Differential analysers (at Manchester University). *The Engineer*, **160**, 19, 26 July, 56–8, 82–4.

Anon. (1935*c*). Differential analyser for the University of Manchester. *Nature*, **135**, 6 April, 533.

Anon. (1935*d*). Electrical calculating machine for simultaneous equations. *Nature*, **135**, 15 May, 851–2.

Anon. (1945*a*). A new mathematics division at the National Physical Laboratory. *Journal of Scientific Instruments*, **22**, 120.

Anon. (1945*b*). Mathematics at the National Physical Laboratory. *Nature*, **155**, 7 April, 431.

Anon. (1946*a*). The National Physical Laboratory at Teddington. *Nature*, **158**, 14 September, 361–3.

Anon. (1946*b*). An Automatic Computing Engine for the National Physical Laboratory. *Nature*, **158**, 7 December, 827.

Anon. (1946*c*). *Description of methods for the numerical solution of simultaneous algebraic linear equations, with examples and details of time taken*, typescript NPL, Mathematics Division Ma 6/05, December 1946.

Anon. (1946*d*). Notes 59. Admirality Computing Service. *Mathematical Tables and other Aids to Computation*, **2**, 15, 188.

Anon. (1946*e*). Computing expert Dr. L. J. Comrie Back From United States. *New Zealand News (London)*, 1 January 1946.

Anon. (1948). The BAASMTC now RSMTC-Final Report of the Committee on the Calculation of Mathematical Tables, Summer 1948. *Mathematical Tables and other Aids to Computation*, **3**, 333–40. Reprinted from *The Advancement of Science*, **5**, 67–72.

Anon. (n.d.). *Memorandum on the centralization of computation in a national mathematical laboratory*, typescript, probably written by J. Todd, D. H. Sadler, and A. Erdélyi in 1944 and presented to the Secretary of the Department of Scientific and Industrial Research, Sir Edward Appleton. Received from J. Todd.

Appleton, E. V. and Weekes, K. (1939). Lunar tides in the upper atmosphere. *Proceedings of the Royal Society*, A, **171**, 171–87.

Astronomer Royal (1939) *Report of the Astronomer Royal to the Board of Visitors to the Royal Observatory, Greenwich 1939*. Public Records Office ADM 190/35, London.

Astronomer Royal (1940). *Report of the Astronomer Royal to the Board of Visitors to the Royal Observatory, Greenwich 1940*. Public Records Office ADM 190/19, London.

Astronomer Royal (1941). *Report of the Astronomer Royal to the Board of Visitors to the Royal Observatory, Greenwich 1941*. Public Records Office ADM 190/20, London.

Astronomer Royal (1942). *Report of the Astronomer Royal to the Board of Visitors to the Royal Observatory, Greenwich 1/5/41–30/4/42*. Public Records Office ADM 190/21, London.

Astronomer Royal (1946). *Report of the Astronomer Royal to the Board of Visitors to the Royal Observatory, Greenwich 1945–1946*. Public Records Office ADM 190/22, London.

BAASMTC (1935). *Mathematical tables* Vol. V, *Factor tables* (ed. L. J. Comrie, J. Peters, A. Lodge, E. J. Ternworth, and E. Gifford). Cambridge University Press, Cambridge.

BAASMTC (1937). *Mathematical tables*, Vol. VI, *Bessel functions*, Part I, *Functions of orders zero and unity* (ed. J. Henderson and L. J. Comrie). Cambridge University Press, Cambridge.

BAASMTC (1946). *Mathematical tables*, Part-Vol. A, *Legendre polynomials* (ed. L. J. Comrie and A. J. Thompson). Cambridge University Press, Cambridge.

BAASMTC (1952). *Mathematical tables*, Vol. X, *Bessel function*, Part II, *Functions of positive integer orders 2 to 20* (W. G. Bickley, L. J. Comrie, J. C. P. Miller, D. H. Sadler and A. J. Thompson). Cambridge University Press, Cambridge.

Babbage, C. (1889). *Tables of logarithms of the natural numbers from 1 to 108,000*. London.

Barlow, P. (1814). *New mathematical tables*. G. and S. Robinson, London. 2nd edn, (ed. A. De Morgan) (1840). 3rd edn, (ed. L. J. Comrie) (1930), *Barlow's Tables of Squares, Roots and Reciprocals of all Integer Numbers up to 10,000*, Spon, London, 4th edn, (ed. L. J. Comrie) (1941).

Bausinger, J. and Peters, J. (1910). *Logarithmic – trigonometrical tables with eight decimal places, containing the logarithms of numbers from 1 to 200,000 . . .*, 2 vols. Wilhelm Engleman, Leipzig.

Baxendall, D. (1926). *Calculating machines and instruments : catalogue of the collections in the Science Museum*, rev. and updated Jane Pugh (1975). Science Museum, London.

Baxendall, D. (1929*a*). Calculating machines. *Encyclopaedia Britannica*, 14th edn, Vol. 4, pp. 548–53.

Baxendall, D. (1929*b*). Mathematical instruments. *Encyclopaedia Britannica*, 14th edn, Vol. 15, pp. 69–72.

Beevers, C. A. (1939). A machine for the rapid summation of Fourier series. *Proceedings of the Physical Society*, **51**, 4, 660–3, discussion 664–7.

Bell, H. and Milne, J. R. (1914). A working list of mathematical tables. In Horsburgh 1914, pp. 47–60.

Blater, J. (1888). *Table of quarter-squares of all whole numbers from 1 to 200,000*. Trübner, London.

Boccardi, G. (1902). *Guide du calculateur (astronomie, geodosie, navigation, etc.)*. Hermann, Paris.

Boys, C. V. (1885). Slide rule. *Nature*, **32**, 627–29.

Boys, C. V. (1901). The comptometer. *Nature*, **64**, 265–8.

Briggs, H. (1617). *Logarithmorum chilias prima*. London.
Briggs, H. (1624). *Arithmetica logarithmica*. London.
British Astronomical Association (1920). Suggested formation of a computing section. *Journal of the British Astronomical Association*, **30**, 9, 313.
British Astronomical Association (1921*a*). Report of annual meeting. *Journal of the British Astronomical Association*, **31**, 16.
British Astronomical Association (1921*b*). Computing section. *Journal of the British Astronomical Association*, **31**, 52.
British Astronomical Association (1921*c*). Computing section. *Journal of the British Astronomical Association*, **31**, 370–1.
British Tabulating Machine Co. Ltd. (1936–7). Minute books, No. 2740 10/11/36, No. 2694 14/9/37, No. 2945 12/10/37. ICL Archive communicated by M. Campbell-Kelly.
British Tabulating Machine Co. Ltd. (1938). *The Tabulator: Monthly Bulletin of the British Tabulating Machine Co. Ltd.*, **2**, 3 (March).
Bromley, A. (1984). Analog computing devices. Unpublished draft. 7 March 1984.
Brown, E. W. (1919). *Tables of the motion of the Moon*, 3 vols. Yale University Press, New Haven.
Buckingham, R. A. and Massey, H. S. W. (1942). The scattering of neutrons by deutrons and the nature of nuclear forces. *Proceedings of the Royal Society*, A, **179**, 123–51.
Bullard, E. C. and Moon, P. B. (1932). A mechanical method for the solution of second-order differential equations. *Philosophical Magazine*, **13**, 546–52.
Burrough, S. M. (1983). Private communication to M. Croarken, 21 February 1983.
Bush, V. (1931). The differential analyser. A new machine for solving differential equations. *Journal of the Franklin Institute*, **212**, 447–88.
Cajori, F. (1909). *History of the logarithmic slide rule and allied instruments*. A. Constable & Co., London.
Callender, A., Hartree, D. R., and Porter, A. (1936). Time-lags in a control system. *Philosophical Transactions*, A, **235**, 415–44.
Callet, F. (1795). *Tables portatives de logarithmes. . . .* F. Didot, Paris.
Cambridge Scientific Instrument Company Ltd. and Mallock, R. R. M. (1937). Legal agreement typescript dated 12 June 1937. C. S. I. Co. Archive in Cambridge University Library, quoted from with permission from the Syndices of Cambridge University Library.
Campbell-Kelly, M. (1980*a*). Programming the EDSAC: early programming activity at the University of Cambridge. *Annals of the History of Computing*, **2**, 1, 7–36.
Campbell-Kelly, M. (1980*b*). Programming the Mark I: early programming activity at the University of Manchester. *Annals of the History of Computing*, **2**, 2, 130–68.
Campbell-Kelly, M. (1980*c*). *Foundations of computer programming in Britain (1945–1955)*. Unpublished Ph.D. thesis.
Campbell-Kelly, M. (1981). Programming the Pilot ACE: early programming activity at the National Physical Laboratory. *Annals of the History of Computing*, **3**, 2, 133–62.
Campbell-Kelly, M. (1982). The development of computer programming in Britain (1945–1955). *Annals of the History of Computing*, **4**, 2, 121–39.
Carroll, J. A. (1933). Letter to Comrie, 16 April 1933. Royal Greenwich Observatory Archives RGO 16/14.

Carroll, J. A. (1943). Admiralty Computing Service. S.R.E.MA/1../3, 12 March 1943, typescript.

Carse, G. A. and Urquhart, J. (1914). Planimeters. In Horsburgh 1914 pp. 190–206.

Carter, A. E. and Sadler, D. H. (1948). The application of the National Accounting Machine to the solution of first order differential equations. *Quarterly Journal of Mechanics and Applied Mathematics*, **1**, 433–41.

Chambers, C. C. (1947). *Theory and techniques for the design of electronic digital computers. Lectures delivered 8 July–31 August 1946*. Moore School of Engineering, University of Pennsylvania, Philadelphia.

Chambers, W. & R. (1844). *Mathematical tables . . .*, (ed. Andrew Bell). W. & R. Chambers, Edinburgh.

Comrie, L. J. (1921). First report of the Computing Section. *Memoirs of the British Astronomical Association*, **24**, Reports of the Observing Sections, 1–28. Bound volume is dated 1924; report is dated 1921.

Comrie, L. J. (1925*a*). The application of calculating machines to astronomical computing. *Popular Astronomy*, **33**, 243–6.

Comrie, L. J. (1925*b*). Logarithmic and trigonometrical tables. *Monthly notices, Royal Astronomical Society*, **85**, 326–88.

Comrie, L. J. (1927). Professional work in Departments of State: XI His Majesty's Nautical Almanac Office. *State Service*, **7**, June 1927, No. 3, 52–3.

Comrie, L. J. (1928*a*). On the application of the Brunsviga-Dupla calculating machine to double summation with finite differences. *Monthly Notices, Royal Astronomical Society*, **88**, 5, 447–59.

Comrie, L. J. (1928*b*). On the construction of tables by interpolation. *Monthly Notices, Royal Astronomical Society*, **88**, 506–23.

Comrie, L. J. (1928*c*). Unpublished letter to Miss M. E. Williams, 18 December 1928. Royal Greenwich Observatory Archives RGO 16/14.

Comrie, L. J. (1928*d*). The use of office equipment: I Division by non-automatic calculating machine. *Office Management and the Office Manager*, 10 September 1928, 17.

Comrie, L. J. (1928*e*). The use of office equipment: V Division by automatic calculating machine. *Office Management and the Office Manager*, 8 October 1928, 6.

Comrie, L. J. (1929*a*). Appendix: tables for interpolation to tenths and fifths by the end-figure process. *Nautical almanac 1931*, pp. 828–35 + attached tables pp. 836–59. HMSO, London.

Comrie, L. J. (1929*b*). German calculating machine enterprise. *Transactions Office Machinery Users' Association 1928–29*.

Comrie, L. J. (1929*c*). Ephemerides, calculating machines. *Encyclopaedia Britannica*, 14th edn, Vol. 8, p.640.

Comrie, L. J. (1930*a*). The Hollerith and Powers tabulating machinery. *Transactions Office Machinery Users' Association 1929–1930*.

Comrie, L. J. (1930*b*). Tabulating machines; adding machines; calculating machines. In *Business encyclopaedia*. Caxton Publishing Co., London.

Comrie, L. J. (1932*a*). Modern Babbage machines. *Bulletin of the Office Machinery User's Association, 1931–1932*, 29pp.

Comrie, L. J. (1932*b*). Report of the Council to the Hundred and Twelfth Annual General Meeting. Mathematical tables. *Monthly Notices, Royal Astronomical Society*, **92**, 338–47.

Comrie, L. J. (1932*c*). The application of the Hollerith tabulating machine to Brown's tables of the Moon. *Monthly Notices, Royal Astronomical Society*, **92**, 694–707.
Comrie, L. J. (1932*d*). The Nautical Almanac Office Burroughs Machine. *Monthly Notices, Royal Astronomical Society*, **92**, 523–41.
Comrie, L. J. (1932*e*). The computation of solar eclipses. *Monthly Notices, Royal Astronomical Society*, **93**, 175.
Comrie, L. J. (1932*f*). The total solar eclipse of 1940, October 1. *Monthly Notices, Royal Astronomical Society*, **93**, 181.
Comrie, L. J. (1933*a*). Computing the nautical almanac. *Nautical Magazine*, July 1933, 33–48.
Comrie, L. J. (1933*b*). *The Hollerith and Powers tabulating machines*. Printed for private circulation, London.
Comrie, L. J. (1935). Adding machines (Vol. I, pp. 22–4); Calculating machines (Vol. I, pp.315–19). In *Hutchinson's technical and scientific encyclopaedia* (ed. C. F. Tweeny and I. P. Shirshov), 3 vols. Hutchinsons & Co. Ltd., London.
Comrie, L. J. (1936*a*). *Interpolation and allied tables*, repr. of Comrie 1929*a*. HMSO, London.
Comrie, L. J. (1936*b*). Inverse interpolation and scientific applications of the National Accounting Machine. *Supplement to the Journal Royal Statistical Society*, **3**, 2, 87–114.
Comrie, L. J. (1937). The application of the Brunsviga Twin 13z calculating machine to the Hartmann formula for the reduction of prismatic spectrograms. *The Observatory*, **60**, 70–3.
Comrie, L. J. (1938*a*). Calculating machines, Appendix III. In *Statistics in theory and practice*, (ed. L. R. Connor), pp. 349–71. Pitman, London.
Comrie, L. J. (1938*b*). *On the application of the Brunsviga Twin 13z calculating machine to artillery survey* (2nd Edition 1940. *On the application . . . to survey problems*). Scientific Computing Service Ltd., London.
Comrie, L. J. (compiled 1938*c*). *Hughes' tables for sea and air navigation*. Henry Hughes, London.
Comrie, L. J. (1938*d*). Unpublished letter to A. T. Doodson, 11 February 1938. Material left by A. Fletcher in possession of Professor Delves, Liverpool University, Item 35.
Comrie, L. J. (1938*e*). Unpublished letter to A. G. Davin, 12 February 1938. Material left by A. Fletcher in possession of Professor Delves, Liverpool University. Item 35.
Comrie, L. J. (1939). A machine for the rapid summation of Fourier series (discussion following a paper by C. A. Beevers). *Proceedings of the Physical Society*, **51**, 4, 660–7.
Comrie, L. J. (1942). *The Twin Marchant calculating machine and its application to survey problems*, repr. 1943, 1945. Scientific Computing Service Ltd., London.
Comrie, L. J. (1944). Recent progress in scientific computing. *Journal Scientific Instruments*, **21**, August, 129–35.
Comrie, L. J. (1946*a*). Babbage's dream come true. *Nature*, **158**, October, 567–8.
Comrie, L. J. (1946*b*). The application of commercial calculating machines to scientific computation. *Mathematical tables and other aids to computation*, **2**, 149–59.
Comrie, L. J. (1947). Calculating—past, present and future. *Future, the Magazine of Industry, Government, Science, Arts*, **1**, Overseas Issue, 61–9.
Comrie, L. J. (1948*a*). *Chambers's six-figure mathematical tables*, Vol. I, *Logarithmic values*. W. & R. Chambers Ltd., Edinburgh.
Comrie, L. J. (1948*b*). War work of the SCS. Transcript of a radio broadcast by L. J.

Comrie. In *New Zealand Revisited and Australia Visited* (W. E. Block, L. J. Comrie, and P. B. Comrie). Typescript copied for private circulation. Scientific Computing Service Ltd., London.

Comrie, L. J. (1949). *Chambers's six-figure mathematical tables*, Vol. II, *Natural values*. W. & R. Chambers Ltd., Edinburgh.

Comrie, L. J. (1950). *Chambers's shorter six-figure mathematical tables*. W. & R. Chambers Ltd., Edinburgh and London.

Comrie, L. J., Hey, G. B., and Hudson, H. G. (1937). The application of Hollerith equipment to an agricultural investigation. *Supplement to the Journal Royal Statistical Society*, **4**, 2, 210–24.

Comrie, L. J. and Levin, A. E. (1921). Eclipse of Rhea by the shadow of Titan. *Monthly Notices, Royal Astronomical Society*, **81**, 486–7.

Comrie, L. J. and Milne-Thomson, L. M. (1931). *Standard four-figure mathematical tables, edn A, with positive characteristics in the logarithms*; *edn B, with negative characteristics in the logarithms*. MacMillan, London.

Comrie, L. J. and Peters, J. T. (1939). *Achtstellige Tafel der trigometrischen Funktionen fur jede Sexagesimalsekunde des Quadranten*. Landesaufnahme, Berlin.

Comrie, P. B. (1984). Private communication to M. Croarken, 13 August 1984.

Conolly, B. W. (1947). The Admiralty Computing Service, *Journal of The Royal Naval Scientific Service*, **2**, 2 (March 1947), 85–8. Public Records Office ADM 206/10.

Copple, C., Hartree, D. R., Porter, A., and Tyson, H. (1939). The evaluation of transient temperature distributions in a dielectric in an alternating field. *Journal Institute of Electrical Engineers*, **85**, 56–65.

Cowling, T. G. (1971). Sidney Chapman 1888–1970. *Biographical Memoirs of Fellows of the Royal Society*, **17**, 53–90.

Cox, E. G., Gross, L., and Jeffrey, G. A. (1947). Uses of punched card tabulating machines for crystallographic Fourier synthesis. *Nature*, **159**, 433–4.

Crank, J. (1947). *Differential analysers*. Longmans, London.

Crank, J., Hartree, D. R., Ingham, J., and Sloane, R. W. (1939). Distribution of potential in cylindrical thermionic valves. *Proceedings of the Physical Society*, **51**, 952–971.

Crelle, A. L. (1820). *Rechentaflen*. G. Reimer, Berlin. First English edn 1897, *Dr. A. L. Crelle's calculating tables* (rev. C. Bremiker). David Nutt, London.

Curtis, A. R. (1983). Personal communication to M. Croarken, 10 May 1983.

Curtis, J. M. (1983). Personal communication to M. Croarken, 18 August 1983.

Darwin, C. G. (1958*a*). Obituaries: J. R. Womersley. *Nature*, **181**, 3 May 1958, 1240.

Darwin, C. G. (1958*b*). Douglas Rayner Hartree, *Biographical Memoirs of Fellows of the Royal Society*, **4**, 103–16.

Desborough, W. (1932–4). Correspondence with Comrie. Royal Greenwich Observatory Archives RGO 16/15.

d'Ocagne, M. (1884). Procède nouveau de calcul graphique. *Anns. des Ponts et Chassées*, **8**, 531–40.

d'Ocagne, M. (1920). Histoire des machines à calculer. *Bulletin de la Société d'Encouragement pour l'Industrie Nationale*, **132**, 554–69.

d'Ocagne, M. (1922). Vue d'ensemble sur les machines à calculer. *Bulletin des Sciences Mathématiques*, 2è séries, **46**, 102–44.

d'Ocagne, M. (1928). *Le calcul simplifié*, 3rd edn. Gauthier-Villars, Paris. Translated

and reprinted 1986, Charles Babbage Institute Series for the History of Computing Vol. 11. MIT Press and Tomash Publishers, Los Angeles.
Drath, P. (1973). *The relationship between science and technology : university research and the computer industry 1945–1962*. Unpublished Ph.D. thesis. University of Manchester.
DSIR (1942–7). *DSIR Advisory Council minutes*. Public Records Office DSIR 1/10, London.
DSIR (1944) *Report of Interdepartmental Technical Committee on a proposed Central Mathematics Station*. Public Records Office DSIR 2/204, London.
DSIR (1946). *Advisory Council minutes 1945–1946*, Minute 143, 1945–6, 10 July 1946. Public Record Office DSIR 1/10, London.
Duffield, W. W. (1895–6). Appendix xii, Tables 1 to 100,000. In *Report of the U.S. coast and geodetic survey*. Washington.
Eames, C. and Eames, R. (office of) (1973). *A computer perspective* (ed. G. Fleck). Harvard University Press, Cambridge, Mass.
Engineering Research Associates (1950). *High speed computing devices* (ed. W. W. Stifler). McGraw-Hill, New York. Reprinted 1983, The Charles Babbage Institute Reprint Series for the History of Computing, Vol. 4. Tomash Publishers, Los Angeles.
Erdélyi, A. and Todd, J. (1946). Advanced instruction in practical mathematics. *Nature*, **158**, 690–2.
Ershov, A. P. (1980). *The British lectures*. Heydon & Son, London.
Evesham, H. A. (1985). Nomography – its rise and fall. Typescript of article to be published in *Annals of the History of Computing*.
Fairthorne, R. A. (1983*a*). Private communication to M. Croarken, 14 March 1983.
Fairthorne, R. A. (1983*b*). Private communication to M. Croarken, 21 March 1983.
Fairthorne, R. A. (1983*c*). Private communication to M. Croarken, 7 September 1983.
Fairthorne, R. A., Griffith, G. J., and Woollett, M. F. C. (1958). *The R.A.E. Hollerith punched card installation, 1945–1952*, UK Aeronautical Research Council, Reports & Memoranda 3084. HMSO, London.
Fairthorne, R. A. and Miller, J. C. P. (1949). Hilbert's double series theorem on principle latent roots of the resulting matrix. *Mathematical Tables and other Aids to Computation*, **3**, 399–400.
Fletcher, A., Miller, J. C. P., and Rosenhead, L. (1946). *An index of mathematical tables*. Scientific Computing Service Ltd, London.
Fox, L. (probably) (1951). The work of Mathematics Division. Typescript Ma 19/2/03. 21 May 1951.
Fox, L. (1952). The work of Mathematics Division. Typescript Ma 19/2/04. 1 May, 1952.
Fox, L., Huskey, H. D., and Wilkinson, J. H. (1948). Notes on the solution of algebraic linear simultaneous equations. *Quarterly Journal of Mechanics and Applied Mathematics*, **1**, 2, 149–73.
Frame, J. S. (1945). Machines for solving algebraic equations. *Mathematical Tables and other Aids to Computation*, **1**, 337–53.
Frazer, R. A. (1940). *On the possible application of Hollerith machines to mathematics*, ARC Oscillation Sub-Committee, 19th Meeting 4 October 1940. No. 4712 0.194.
Gifford, E. (1914). *Natural sines*. Heyward, Manchester.
Gilles, D. C. (1983). Private communication in M. Croarken, 13 October 1983.

Gittus, E. (1983). Private communication to M. Croarken, 21 November 1983.

Glashier, J. W. L. (1873). Report of the Committee. . . . on Mathematical Tables. *Report of the British Association for the Advancement of Science*, **8**, (Bradford), 1–175.

Glashier, J. W. L. (1911*a*). Logarithm. *Encyclopaedia Britannica*, 11th edn. Vol. 16, pp.868–77.

Glashier, J. W. L. (1911*b*). Tables, mathematical. *Encyclopaedia Britannica*, 11th edn, Vol. 26, pp.325–36.

Glennie, A. E. (1978). Letter to M. Campbell-Kelly, 17th November 1978.

Goldstine, H. H. (1972). *The computer from Pascal to von Neumann*. Princeton University Press, New Jersey.

Goode, I. (1983). Private communication to M. Croarken, 10 May 1983.

Goodwin, E. T. (1952). *Reorganisation of Mathematics Division*. Typescript Ma 33/01. 31 January 1952.

Goodwin, E. T. (1955). Mathematics Division. *NPL news*, 15 August 1955, No. 1000000 (=64), 1–10 (=1–2).

Goodwin, E. T. (1984). Private communication to M. Croarken, 4 December 1984.

Grace, S. F. (1934). Correspondence with L. J. Comrie. Royal Greenwich Observatory Archives RGO 16/14.

Greaves, W. H. M. (1953). Obituary notices: Leslie John Comrie. *Monthly Notices, Royal Astronomical Society*, **113**, 3, 294–304.

Hailstone, J. E. (1958). *The use of the Hollerith Type 555 electronic calculator for scientific computation*. British Tabulating Machine Company Ltd., with permission from AERE, Harwell.

Hailstone, J. E. (1983*a*). Private communication to M. Croarken, 18 June 1983.

Hailstone, J. E. (1983*b*). Private communication to M. Croarken, 26 June 1983.

Hantschl, J. (1827). *Logarithmisch-trigonometrisches handbuch*, Vienna.

Hartley, H. O. (1946). The application of some commercial calculating machines to certain statistical calculations. *Journal Royal Statistical Society (Suppl.)*, **8**, 2, 154–183.

Hartree, D. R. (1929). Letter to L. J. Comrie, 26 November 1929. Royal Greenwich Observatory Archives RGO 16/15.

Hartree, D. R. (1934). Approximate wave functions and atomic field of mercury. *Physical Review*, **46**, 738–43.

Hartree, D. R. (1935). The differential analyser. *Nature*, **135**, 940–3.

Hartree, D. R. (1936). The application of the differential analyser to the solution of partial differential equations. *Report of the International Congress of Mathematicians*. Oslo.

Hartree, D. R. (1938). The mechanical integration of differential equations. *Mathematical Gazette*, **22**, 342–64.

Hartree, D. R. (1940*a*). A great calculating machine. *Proceedings of the Royal Institution*, **31**, 151–70.

Hartree, D. R. (1940*b*). The Bush differential analyser and its application. *Nature*, **146**, 319–23.

Hartree, D. R. (1943). The thirty fourth Kelvin lecture. Mechanical integration in electrical problems. *Journal of the Institute of Electrical Engineers*, **90**, 435–42.

Hartree, D. R. (ed.) (1947*a*). *The differential analyser*, Ministry of Supply Permanent Records of Research and Development, Monograph No. 17.502.

Hartree, D. R. (1947*b*). *Calculating machines – recent and prospective developments and their impact on mathematical physics*, Inaugural Lecture. Cambridge University Press, Cambridge.

Hartree, D. R. (1950). *Calculating instruments and machines*. Cambridge University Press, Cambridge.

Hartree, D. R. (1952). *Numerical analysis*, 2nd edn. 1958. Clarendon Press, Oxford.

Hartree, D. R. and Ingham, J. (1938). Note on the application of the differential analyser to the calculation of train running times. *Memoirs Manchester Literary and Philosophical Society*, **83**, 1–15.

Hartree, D. R., Myers, D. M., and Porter, A., (1937). The effect of space-charge on secondary current in a triode. *Proceedings of the Royal Society*, A, **158**, 23.

Hartree, D. R. and Nuttall, A. K. (1938). The differential analyser and its applications in electrical engineering. *Journal of the Institute of Electrical Engineers*, **83**, 648.

Hartree, D. R., Nuttall, A. K., and Porter, A. (1936). The response of a non-linear electric circuit to an impulse. *Proceedings of the Cambridge Philosophical Society*, **32**, 304.

Hartree, D. R. and Porter, A. (1935). The construction and operation of a model differential analyser. *Memoirs Manchester Literary and Philosophical Society*, **79**, 51–72.

Hartree, D. R. and Porter, A. (1938). The application of the differential analyser to transients on a distortionless transmission line. *Journal of the Institution Electrical Engineers*, **83**, 648–56.

Hartree, D. R., Porter, A., Callender, A., and Stevenson, A. B. (1937). Time-lag in a control system, II. *Proceedings of the Royal Society*, A, **161**, 460.

Hartree, D. R. and Womersley, J. R. (1937). A method for the numerical or mechanical solution of certain types of partial differential equations. *Proceedings of the Royal Society*, A, **161**, 353–66.

Henrici, O. (1894). Report on planimeters. *Reports of the British Association for the Advancement of Science 1894*, pp.496–523, London.

Henrici, O. (1911). Calculating machines. *Encyclopaedia Britannica*, 11th edn, Vol. 4, pp.972–81.

Herwart ab Hohenburg, J. G. (1610). *Tabulae arithmeticae universales, quarun. . . .* Monachii, Bavariarium.

Hey, G. B. (1983). Private communication to M. Croarken, 22 November 1983.

Hodges, A. (1983). *Alan Turing: the enigma*. Burnett Books, London.

Hodgkin, D. Crowfoot (1949). The X-ray analysis of the structure of penicillin. *Advancement of Science*, **6**, 85–9.

Hodgson, M. L., Clews, C. J. B., and Cochran, W. (1949). A punched card modification of the Beevers-Lipson method of Fourier synthesis. *Acta Crystallographa*, **2**, 113–16.

Hollingdale, S. H. (1983*a*). Private communication to M. Croarken, 11 March 1983.

Hollingdale, S. H. (1983*b*). Private communication to M. Croarken, 21 November 1983.

Hooper, F. J. (1983). Private communication to M. Croarken, 28 October 1983.

Horsburgh, E. M. (ed.) (1914). *Handbook of the Napier tercentenary celebration*, or *Modern instruments and methods of calculation*. Bell, London. Reprinted 1982, Charles Babbages Institute Reprint Series for the History of Computing, Vol. 3. Tomash Publishers, Los Angeles.

Horsburgh, E. M. (1922). Calculating machines. In *Dictionary of applied physics* (ed. Sir Richard Glazebrook), Vol. 3, pp.193–201. Macmillan & Co., London.

Hoskin, N. E. (1978). Letter to M. Campbell-Kelly, 20 December 1978.

Howlett, J. (1980). Computing at Harwell. In *25 Years of Theoretical Physics 1954–1979*, Chapter 21. Report TP.843, Theoretical Physics Division, AERE, Harwell, Oxfordshire, March 1980.

Hudson, T. C. (1914). HM Nautical Almanac Office anti-differencing machine. In Horsburgh 1914, pp.127–31.

Huskey, H. D. (1984). From ACE to the G-15. *Annals of the History of Computing*, **6**, 4, 350–70.

Jacob, L. (1911). *Le calcul mècanique*. Encyclopedie Scientifique Series. Octave Doin et Fils, Paris.

Jones, R. V. (1978). *Most secret war*. Hamish Hamilton, London.

Keuffel, A. W. (1971). Slide rule. *Encyclopaedia Britannica*, 1971 edn, Vol. 20, pp.653–5.

Kilburn, T. and Piggott, L. S. (1978). Frederic Calland Williams 1911–1977. *Biographical Memoirs of Fellows of the Royal Society*, **24**, 583–604.

Knott, C. G. (ed.) (1915). *Napier tercentenary memorial volume*. Royal Society of Edinburgh/Longmans, Green & Co., London.

Lavington, S. H. (1975). *A history of Manchester computers*. NCC Publications, Manchester.

Lavington, S. (1980). *Early British computers*. Manchester University Press, Manchester.

Lennard-Jones, J. (1937). Untitled typescript 2pp, dated 2 April 1937, Churchill College Archives, University of Cambridge, Item LEJO 6. Quoted with the permission of The Master, Fellows and Scholars of Churchill College of the University of Cambridge.

Lennard-Jones, J. (1942). Progress of work journals, Churchill College Archives, University of Cambridge, Item LEJO 29. 2 April–26 September 1942.

Lennard-Jones, J. (1943). Daily journals, Churchill College Archives, University of Cambridge, Item LEJO 24, (5) 31/3/43; (8) 27/5/43.

Lennard-Jones, J. (1944–6). Personal journals 28 September 1944–20 March 1946, Churchill College Archives, University of Cambridge, Item LEJO 26.

Lennard-Jones, J. (*c.* 1946). Manuscript curriculum vitae, Churchill College Archives, University of Cambridge, Item LEJO 8.

Lennard-Jones, J., Wilkes, M. V., and Bratt, J. B. (1939). The design of a small differential analyser. *Proceedings of the Cambridge Philosophical Society*, **35**, 485–93.

Lilley, S. Mathematical machines. *Nature*, **149**, 25 April, 462–5.

Lipson, H. and Beevers, C. A. (1936). An improved numerical method of two-dimensional Fourier synthesis. *Proceedings of the Physical Society*, **48**, 772–80.

Livesley, R. K. (1979). Letter to M. Campbell-Kelly, 26 February 1979.

Lohse, O. W. (1909). *Tafeln fur Numerisches Rechnen mit Maschinen*. Engelmann, Leipzig.

Martin, E. (1925). *The calculating machines and the history of their development*, Vol. I, *Calculating machines with automatic tens-carry*. Translation by KAT/HMcD 4/29/46. Johannes Meyer, Pappenheim.

Mallock, R. R. M. (1933). An electrical calculating machine. *Proceedings of the Royal Society*, A, **140**, 457–83.

Massey, H. S. W. (1952). L. J. Comrie. *Obituary Notices of Fellows of the Royal Society*, **8**, 97–107.
Massey, H. S. W., Wylie, J., Buckingham, R. A. and Sullivan, R. (1938). Small scale differential analyser: its construction and operation. *Proceedings of the Royal Irish Academy*, A, **45**, 1–21.
Mathias, P. (1983). *The first industrial revolution: an economic history of Britain 1700–1914*, 2nd edn. Methuen, London.
McCarthy, J. H. (1924). *The American digest of business machines: a compendium of makes and models with specifications and principles of operation described, including used machine evaluations*. American Exchange Service, Chicago.
Michel, J. G. L. (1955). The mechanical differential analyser: recent developments and improvements. *Journées Internationales du Calcul Analogique*, September.
Miller, J. C. P. (1954). *Royal Society mathematical tables series*, Vol. 3, *Binomial coefficients*. Cambridge University Press for the Royal Society, London.
Miller, J. C. P. (1979). Private communication to M. Campbell-Kelly, June 1979.
Ministry of Supply (n.d.) Ministry of Supply Advisory Council on Scientific Research and Technical Development; Ballistics Committee. Public Records Office WO 195/5015, London.
Mott, N. F. (1955). John Edward Lennard-Jones. *Biographical Memoirs of Fellows of the Royal Society*, **1**, 175–84.
Murray, F. J. (1948). *The theory of mathematical machines*. Kings-Crown Press, New York.
Napier, J. (1614). *Mirific logarithmorum canonis descriptio*. Edinburgh.
Nautical Almanac Office (1961). *Explanatory supplement to the astronomical ephemeris and the American ephemeris and nautical almanac*. HMSO, London.
Newman, M. H. A. (1946). Letter to John von Neumann, 8 February 1946. Quoted in Goldstine 1972, p. 249.
Newman, M. H. A. (1948). *A status report on the Royal Society Computing Machinery Laboratory*, Minutes of the Senate Sub-Committee of Manchester University. Arrangements for the Royal Society Computing Machine Laboratory. Communication to M. Croarken by D.B.G. Edwards, 18 April 1988.
Newman, M. H. A. (1955). Alan Mathison Turing. *Biographical Memoirs of Fellows of the Royal Society*, **1**, 253–63.
NPL (1942–7). Executive Committee minutes and reports, 1945–7. Public Record Office DSIR 10/366, London.
NPL (1949*a*). *Report of the NPL 1946*. DSIR, HMSO, London.
NPL (1949*b*). *Report of the NPL 1947*. DSIR, HMSO, London.
NPL (1950*a*). *Report of the NPL 1948*. DSIR, HMSO, London.
NPL (1950*b*). *Report of the NPL 1949*. DSIR, HMSO, London.
NPL (1951). *Report of the NPL 1950*. DSIR, HMSO, London.
NPL (1952*a*). *Report of the NPL 1940–1945*. DSIR, HMSO, London.
NPL (1952*b*). *Report of the NPL 1951*. DSIR, HMSO, London.
NPL (1953). *Report of the NPL 1952*. DSIR, HMSO, London.
NPL (1954*a*). *Automatic digital computation: proceedings of a symposium held at the National Physical Laboratory on 25–28 March 1953*. HMSO, London.
NPL (1954*b*). *Report of the NPL 1953*. DSIR, HMSO, London.
NPL (1955). *Report of the NPL 1954*. DSIR, HMSO, London.

NPL (1956). *Report of the NPL 1955*. DSIR, HMSO, London.

NPL (1957). *Report of the NPL 1956*. DSIR, HMSO, London.

NPL (1958). *Report of the NPL 1957*. DSIR, HMSO, London.

NPL (1959). *Report of the NPL 1958*. DSIR, HMSO, London.

Ordnance Board (1942). *Ordnance Board Proceedings 18,801–19,271*, No. 18,857 Ballistics, External. Public Records Office SUPP 6/405, London.

Ordnance Board (1945). *Supplement to the Report of the President of the Ordnance Board 1/12/41–31/12/45 Part 2*. Public Records Office AVIA 44/58, London.

Pearson, E. S. (1936). Karl Pearson: some aspects of his life and work, Part I, 1857–1906. *Biometrika*, **28**, 193–257.

Pearson, E. S. (1937). Karl Pearson: some aspects of his life and work, Part II, 1906–1936. *Biometrika*, **29**, 161–248.

Pearson, K. (1904). Letter to Dr Foster of the Drapers' Company. In E. S. Pearson 1936, Appendix II, pp. 254–7.

Pearson, K. (1918). War work of the Biometrics Laboratory. In E. S. Pearson 1937, Appendix III, pp. 241–4.

Pearson, K. (1919). Prefatory note. In *Tracts for computers No. 1. Tables of the digamma and trigamma function*. Eleanor Pairman, Department of Applied Statistics, University College, London. Cambridge University Press, London.

Petherick, E. J. (1949). *The RAE sequence controlled calculator*, RAE Technical Note M.S. 1 October 1949.

Pitiscus (1613). *Thesaurus mathematicus*. . . . Frankfurt.

Plummer, H. C. (1912). Logarithmic tables. *Monthly Notices, Royal Astronomical Society*, **72**, 340–2.

Porter, J. G. (1951). Punched card machines at Herstmonceaux. *Journal British Astronomical Association*, **61**, 7, 185–9.

Porter, J. G. (1953). Obituary—Leslie John Comrie. *Observatory*, **71**, 860, 24–36.

Pugsley, A. G. (1940). *Calculating machines for evaluating determinants*, Report no. AD.3155 Jan 1940. RAE, Farnborough.

Pugsley, A. G. (1983*a*). Private communication to M. Croarken, 23 March 1983.

Pugsley, A. G. (1983*c*). Private communication to M. Croarken, 5 October 1983.

Pugsley, A. G. (1984). Private communication to M. Croarken, 20 February 1984.

Pyatt, E. (1983). *The National Physical Laboratory—a history*. Adam Hilger Ltd., Bristol.

Randell, B. (1982). *The origins of digital computers: selected papers*, 3rd edn. Springer-Verlag, Berlin.

Rees, M. (1982). The computing program of the Office of Naval Research 1946–1953. *Annals of the History of Computing*, **4**, 2, 102–20.

Rheticus, G. J. (1596). *Opus Palatinium de triangulus*. Valantine Otho, Neustadl.

Rigg, F. A. (1983). Private communication to M. Croarken, 17 July 1983.

Robb, A. M. (1914). The use of mechanical integrating machines in naval architecture. In Horsburgh 1914, pp. 206–17.

Rosenhead, L. (1954*a*). Society and the calculating machine. *Advancement of Science*, **10**, 421–433.

Rosenhead, L. (1954*b*). *Report on National Class 3000 accounting machine now on indefinite loan from the Royal Society to the Department of Applied Mathematics, The University, Liverpool*. Typescript, Royal Society Library Mathematical Tables Committee Report MT/21(54).

Royal Society (1946). *Royal Society Minutes of Council, XVII*, Minute 5, 14 February; Minute 9, 7 March; Minute 9, 11 April; Minute 8, 16 May; Minute 7, 13 June; Minute 7, 11 July.

Royal Society Mathematical Tables Committee (1954). Bessel function zeros. *Committee progress report No. 7*, March 1954.

Sadler, D. H. (1941). Correspondence between Sadler and A. A. Moss, Assistant Director of Research, Ministry of Supply, 7 May 1941, 12 May 1941, 8 July 1941. Royal Greenwich Observatory Archives RGO 16/17.

Sadler, D. H. (1984*a*). Private communication with M. Croarken, 4 December 1984.

Sadler, D. H. (1984*b*). Private communication with M. Croarken, 23 December 1984.

Sadler, D. H. and Todd, J. (1946). Mathematics in government service and industry: some deductions from the war time experience of the Admiralty Computing Service. *Nature*, **157**, 571–3.

Sadler, D. H. and Todd, J. (1947). Admiralty Computing Service. *Mathematical Tables and other Aids to Computation*, **2**, 18, 289–97.

Salomon, B. (1931). Mechanical integrators with holonomic relationships. *Comptes Rendus*, **193**, 830–2.

Salomon, J. (1827). *Logarithmisches Tafeln. . . .* Gerold, Vienna.

Schmidt, R. J. and Fairthorne, R. A. (1940). *Some comments on the use of Hollerith machines for scientific calculations*. Report AD. 3153 September 1940. RAE, Farnborough.

Scientific Computing Service Ltd. (1938*a*). *Scientific Computing Service: list of staff*. Scientific Computing Service Ltd., London.

Scientific Computing Service Ltd. (1938*b*). *Scientific Computing Service: a description of its activities, equipment and staff*. Scientific Computing Service Ltd., London.

Scientific Computing Service Ltd. (1938–73). Minute book.

Scientific Computing Service Ltd. (1945). *Scientific Computing Service: list of staff*. Scientific Computing Service Ltd., London.

Scientific Computing Service Ltd. (1946). *Scientific Computing Service: a description of its activities, equipment and staff*. Scientific Computing Service Ltd., London.

Scientific Computing Service Ltd. (1947). *Scientific Computing Service: list of staff*. Scientific Computing Service Ltd., London.

Scientific Computing Service Ltd. (1950*a*). *Scientific Computing Service: a description of its activities, equipment and staff*. Scientific Computing Service Ltd., London.

Scientific Computing Service Ltd. (*c*. 1950*b*). *List of mathematical tables for sale by the Scientific Computing Service Ltd.*, a pamphlet. Scientific Computing Service Ltd., London.

Staff of the Mathematical Services Department (1952). *Report No. MS52: The RAE Hollerith punched card installation*, Internal Report. RAE, Farnborough.

Stokes, G. D. C. (1914). The slide rule. In Horsburgh 1914, pp. 155–80.

Tee, G. J. (1981). Two New Zealand mathematicians. *Proceedings of the First Australian Conference on the History of Mathematics, Monash University, Clayton, Australia. 6 and 7 November 1980* (ed. J. N. Crossley. Department of Mathematics, Monash University, Clayton, Victoria, Australia 3168).

Thoman, F. (1867). *Tables de logarithmes à 27 décimales pour les calcules de précision*. Paris.

Thomson, W. (1881). The tide gauge, tidal harmonic analyser and tide predictor.

Minutes of Proceedings of the Institution of Civil Engineers, *LXV*, Part III, 1 March 1881, pp.2–25; discussion pp.26–64; correspondence pp.65–72.

Thomson, Sir W. (Lord Kelvin) and Tait, P. G. (1890). *Treatise on natural philosophy Part I*, Appendix B'. Cambridge University Press, London. 1912 edn repr 1962 by Dover Publications, New York under the title *Principles of mechanics and dynamics Part I.*

Todd, J. (1974). John von Neumann and the National Accounting Machine. *SIAM Review*, **16**, 4 (October), 526–30.

Todd, J. (1983). Private communication to M. Croarken, 26 July 1983, 28 July 1983.

Tootill, G. C. (1978). Letter to M. Campbell-Kelly, 17 November 1978.

Turck, J. A. V. (1921). *Origin of modern calculating machines*. The Western Society of Engineers, Chicago.

Turing, A. M. (1936). On computable numbers with an application to the Enscheidungsproblem. *Proceedings of the London Mathematical Society*, **2**, 42, 230–67.

Turing, A. M. (1946). *Proposals for development in the Mathematics Division of an Automatic Computing Engine (ACE)*, Report to the NPL Executive Committee, E882 (presented to the Committee February 1946).

University Mathematics Laboratory, Cambridge (1950). *Report of a conference on high speed automatic calculating machines 22–25 June 1949*. Issued by the University Mathematics Laboratory, Cambridge with the assistance of the Ministry of Supply.

University of Cambridge, Faculty Board of Mathematics (1936). *Report of the Faculty Board of Mathematics on the need for a computing laboratory*, typescript Churchill College Archives, University of Cambridge, Item LEJO 6.

University of Cambridge, General Board (1937*a*). Report of the General Board on the establishment of a computing laboratory. Contains reports dated 2 December 1936, 21 February 1936 and 27 January 1937. *Cambridge University Reporter*, 2 February 1937, 625–9.

University of Cambridge, General Board (1937*b*). Report of the General Board on the organization of a mathematical laboratory. Dated 21 April 1937. *Cambridge University Reporter*, 27 April 1937, 917.

University of Cambridge, General Board (1946). Report of the General Board on the organization of the mathematical laboratory. *Cambridge University Reporter*, 22 October 1946, 212–14.

University of Cambridge, Mathematical Laboratory Committee (1947). *Report of the Mathematical Laboratory Committee to the Faculty Board of Mathematics*, typescript University of Cambridge Library, Manuscript Room, Archives. Computing Laboratory UL COMP F1(2). Quoted with permission from the Syndices of Cambridge University Library.

University of Cambridge, Regent House (1937). Acta No. 2. *Cambridge University Reporter*, 23 February 1937, 694.

University of Manchester (1946–52). *Council minutes*. Extracts selected by D. B. G. Edwards communicated to M. Croarken, 18 April 1988.

University of Manchester (1948). *Minutes of the Senate Sub-Committee, arrangements for the Royal Society Computing Machine Laboratory*. Friday 15 October 1948. Communicated by D. B. G. Edwards to M. Croarken, 18 April 1988.

Vega, G. von (1794). *Thesaurus logarithmorum completus. . . .* Weidmann, Leipzig.

Vlacq, A. (1628). *Arithmetica logarithmica, sive logarithmorum chilides. . . .* Gouda.

Voisin, A. (1817). *Tables de multiplications ou logarithmes des nombres entiers depuis 1 jusqu'à 20,000. . . .* Paris.

Vickers, T. (1978). Letter to the Editor, '. . . and the trip that stimulated his enthusiasm'. *Computing*, 2 February 1978, 13.

Vickers, T. (1983). Personal communication to M. Croarken, 4 April 1983.

Vickers, T. (1984). Personal communication to M. Croarken, 29 November 1984.

Vidal, P. C. (1936). Mechanisms for solving systems of linear equations. *Comptes Rendus*, **202**, 1748–51.

von Neumann, J. (1945). First draft of a report on the EDVAC. Contract no. W-670-ORD-492. Moore School of Electrical Engineering, University of Pennsylvania, 30 June 1945.

Walker, P. B., Coblans, H., and Brookes, B. C. (1974). Robert Arthur Fairthorne. *Journal of Documentation*, **30**, 2, 127–39.

Waters, B. (1983). Private communication to M. Croarken, 25 October 1983.

Wheeler, J. (1983*a*). Private communication to M. Croarken, 20 March 1983.

Wheeler, J. (1983*b*). Private communication to m. Croarken, 11 May 1983.

Whittaker, E. T. and Robson, G. (1924). *The calculus of observations: a treatise on numerical mathematics*. Blackie & Son, Glasgow. 4th edn 1944, repr. 1954.

Wilkes, M. V. (1946*a*). Typescript memorandum. Cambridge University Library, Manuscripts Room, Archives. Computer Laboratory UL COMP F1 (1a).

Wilkes, M. V. (1946*b*). *The Functions of the Mathematical Laboratory*. Typescript report to the mathematics faculty board. Cambridge University Library, Manuscripts Room, Archives. Computer Laboratory UL COMP F1 (1b). Quoted with permission from the Syndices of Cambridge University Library.

Wilkes, M. V. (1975). Early computer developments at Cambridge: the EDSAC. *The Radio and Electronic Engineer*, **45**, 7, 332–5.

Wilkes, M. V. (1984*a*). Introduction to D. R. Hartree, *Calculating instruments and machines*. Charles Babbage Institute Reprint Series for the History of Computing, vol. 6. Tomash Publishers, Los Angeles.

Wilkes, M. V. (1984*b*). Private communication with M. Croarken, 4 May 1984.

Wilkes, M. V. (1985). *Memoirs of a computer pioneer*. MIT Press, Cambridge, Mass.

Wilkes, M. V. and Renwick, W. (1949). The EDSAC, an electronic calculating machine. *Journal of Scientific Instruments*, **26**, 385–91.

Wilkinson, J. H. (1971). Some comments from a numerical analyst. *Journal of the Association of Computing Machinery*, **18**, 2, 137–47.

Wilkinson, J. H. (1975). The Pilot ACE at the National Physical Laboratory. *The Radio and Electronic Engineer*, **45**, 7, 336–40.

Wilkinson, J. H. (1980). Turing's work at the NPL and the construction of the Pilot ACE, DEUCE and ACE. In *A history of computing in the twentieth century: a collection of essays* (ed. N. Metropolis, J. Howlett, and G. Rota), pp.101–14. Academic Press, New York.

Williams, C. (1983). Private communication with M. Croarken, 16 March 1983.

Williams, F. C. (1958). *Computing machine. Statement by Professor F. C. Williams on the running costs of the computing machine*. Report sent to P. D. Greenhall, Head of Grants Division, DSIR, 14 May 1948. Held at National Archive for History of Computing, University of Manchester.

Williams, F. C. (1975). Early computers at Manchester University. *The Radio and Electronic Engineer*, **45**, 7, 327–31.
Williams, M. R. (1982). Introduction to 1982 reprint of Horsburgh 1914.
Womersley, J. R. (1949*a*). *Mathematics Division policy : effect of ACE and new differential analyser on organization, staff requirements and accommodation*, typescript 3 May 1949. National Physical Laboratory, Woodger Papers, Science Museum Library, M11.
Womersley, J. R. (1949*b*). *Scientific Computing Service*, typescript 3 May 1949. Internal memo attached to Womersley 1949*a*.
Womersley, J. R. (1949*c*). Letter from J. R. Womersley to Director, NPL, 25 June 1949, *re* Advisory Committee on Calculating Machines.
Woodger, M. (1958). The history and present use of computers at the National Physical Laboratory. *Process Control and Automation*, November, 438–43.
Woodger, M. (1977). *A little light on the prehistory of ACE and EDSAC : extracts from correspondence on file at the NPL*, typescript 1 February 1977. National Physical Laboratory, Woodger papers, Science Museum library, K10.
Woollett, M. F. C. and Simm, J. (1949). *Punched card computation of binomial coefficients over the range* $n=201(1)1000$; $r=2(1)12$. RAE Technical Note MS2. December 1949.
Yates, F. and Rees, D. H. (1958). The use of an electronic computer in research statistics: four years' experience. *Computer Journal*, **1**, 49–58.
Young, A. (1954). Techniques for the summation of products on Hollerith and national Accounting Machines. *Quarterly Journal of Mechanics and Applied Mathematics*, **7**, 136–51.
Yule, G. U. and Filon, L. N. G. (1938). Karl Pearson 1857–1936. *Obituary Notices of Fellows of the Royal Society*, **2**, 70–110.

INDEX

The manufacturer's authorised representative in the EU for product safety is Oxford University Press España S.A. of el Parque Empresarial San Fernando de Henares, Avenida de Castilla, 2 – 28830 Madrid (www.oup.es/en or product.safety@oup.com). OUP España S.A. also acts as importer into Spain of products made by the manufacturer.

www.ingramcontent.com/pod-product-compliance
Ingram Content Group UK Ltd.
Pitfield, Milton Keynes, MK11 3LW, UK
UKHW020048200726
13851UKWH00021B/33

* 9 7 8 0 1 9 8 5 3 7 4 8 9 *